T0233739

Janus-Faced Probability

Paolo Rocchi

Janus-Faced Probability

 Springer

Paolo Rocchi
IBM
Roma
Italy

and

LUISS University
Roma
Italy

ISBN 978-3-319-35595-5   ISBN 978-3-319-04861-1   (eBook)
DOI 10.1007/978-3-319-04861-1
Springer Cham Heidelberg New York Dordrecht London

Printed on acid-free paper

Springer is part of Springer Science+Business Media (www.springer.com)

*"When you travel often,
you will be addicted to it forever.
Our destination is not the place,
but new way to see things."*

Henry Miller

Preface

Paolo Rocchi's book argues that the diversity of the interpretations of probability, in fact, is not a problem crying for immediate resolution, but a very natural situation for a scientist working on concrete applications of probability and statistics. One need not keep to just one concrete interpretation, say to the frequency interpretation. In the debate on interpretations of probability, which has been continuing for centuries, the idea that interpretations can vary depending on the problems under study can be considered as opportunistic. The probabilistic community is sharply divided into camps struggling to justify one or other concrete interpretation. For example, Kolmogorov and Gnedenko actively supported the objective interpretation, and they strongly criticized the subjective approach (however, this did not affect the friendly relations between Kolmogorov and De Finetti). Although both Kolmogorov and von Mises kept to the objective interpretation, they fiercely debated the primacy of frequency or measure-theoretic probability. It is interesting that although Kolmogorov debated von Mises' fundamental frequentism, till his death he remained not completely satisfied by his (Kolmogorov's) measure-theoretic axiomatics. He worked hard to find a mathematically rigorous definition of a random sequence and, in spite of an era of a few brilliant new ideas, such as Kolmogorov's algorithmic complexity, the aforementioned problem is still open.

This book presents an in-depth analysis of the basic interpretations of probability and it can be considered as an attempt to combine them harmoniously on the basis of the pluralist approach: "interpretations can be chosen depending on applications." This is a complex research project and it is far from being complete. Nevertheless, the book can be considered as one of the most important contributions to the analysis of the interpretational problems of probability theory, at least in the last 10–15 years.

Although personally I do not support the author's pluralism (I consider the diversity of interpretations a sign of the deep crises in the foundations of probability as the result of the absence of rigorous mathematical theory of individual random sequence), I have to agree with his observation that if one of the basic interpretations, e.g., frequency or subjective, were wrong, we would have long seen problems with applications.

I also point to the similarity between the present interpretational statuses of probability theory and quantum mechanics. It might be that the interpretational problems of the latter are generated by the interpretational problems of the former.

A collateral message of this book is that developing the pluralist viewpoint on probability will contribute to multidisciplinary applications of probability, but it is not the final word on the problem of interpretations of probability. I hope it will pave the way for many new books to come.

Andrei Khrennikov
Professor of Applied Mathematics,
Intl. Center for Math. Modeling
Linnaeus University
Växjö, Sweden

Contents

Introduction

The English word '*philosopher*' usually denotes a specialist who thinks deeply into things and investigates into broad themes such as the destiny of man, the limit of human knowledge, and so forth.

This book deals with the foundations of the probability calculus, a broad topic indeed, but I feel that the adjective '*philosophical*' is an inappropriate attribute for this work which gives priority to particulars rather than to extensive arguments.

I address some fundamental aspects of the probability using the *analytical approach*, and mean to show how "tiny" mathematical components can suggest innovative answers to the problems long debated.

The first part of the book deals with the frequentist and subjective interpretations of probability. The second part analyzes some aspects of the probability axiomatization.

Part I
On the Meaning of Probability

Chapter 1
Interpretations of Probability

It may be said that the probability calculus is infiltrating almost all the areas of research and common life. Investors forecast the chances of making profits; politicians try to see the orientation of compatriots; sick persons want information about the possibility of them healing from a disease; engineers figure out the state of a construction. Probability is used in analyzing genetics, weather prediction, and a myriad of other everyday events. Probability is the remedy to the ignorance of what will happen next, of what will intervene and pertain to future events. Laymen and scientists, carefree players and serious leaders, criminal and honest men draw attention to statistical and probability values, which, beyond any doubt, can be defined as the most popular mathematical elements in the modern economies.

This great success is somewhat surprising because there is no satisfactory real-world interpretation of the probability concept which statistics is based on. The wide–spreading circulation of statistical data is self evident although the nature of probability still appears rather obscure. The following puzzling question waits for an answer:

What does the term '*probability*' really mean?

People have a prismatic concept in what probability consists of, due to their multi-fold involvement in future and indeterminate occurrences. The Encyclopedia of Philosophy (Edwards 1967) comments on the common sense of the term '*probability*' and itemizes twenty-seven ordinary uses of this word. In general there is a certain fluid relationship between the common use of a word, the scientific realm and philosophy; Eagle (2004) sums up his opinion on this process by means of a graph (Fig. 1.1). Scientists, laymen, and philosophers work together in a rather unofficial way. The first ones provide the interpretation of a word on the basis of explications which stem from folk tradition. Eagle concludes that when a theory disagrees too much with the ordinary usage of the words; it simply will fail.

P. Rocchi, *Janus-Faced Probability*, DOI: 10.1007/978-3-319-04861-1_1,
© Springer International Publishing Switzerland 2014

Fig. 1.1 Relations amongst
concepts

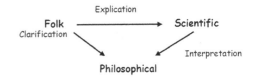

1.1 Variety of Theories

Common life, scientific discoveries and philosophical arguments stimulated
experts who devised the following principal visions of the term *'probability'* in the
span of the last three centuries (Weatherford 1982; Maistrov 1974):

(1) *Classical Interpretation* authored by Pierre Simone de Laplace.
(2) *Frequentist Interpretation* developed in the early twentieth century by the
    Austrian Richard von Mises and others such as the German Hans
    Reichenbach.
(3) *Axiomatic Interpretation* formalized by Andrey Nikolayevich Kolmogorov.
(4) *Subjective Interpretation* elaborated in the same year by the Italian Bruno de
    Finetti and Frank Plumpton Ramsey, but independently of each other.
(5) *Bayesian Interpretation* refers to methods in probability and statistics named
    after the Reverend Thomas Bayes.
(6) *Logic Interpretation* sustained by Rudolf Carnap, John Maynard Keynes, the
    Ludwig Wittgenstein and others.
(7) *Propensity Interpretation* devised by Karl Raimund Popper and shared by
    several followers of Popper.

Perhaps the reader has partial knowledge of the works listed above, and for the
sake of completeness Appendix A recounts the essential traits of theories from
1 to 7. This summary also includes historical notes for the purpose of clarifying
the cultural extraction of the various viewpoints on probability.

Entries 1–7 appear not-trivial although they do not provide the complete
account of the wealth of theoretical constructions which have been put forward to
clarify the essence of probability. The whole spectrum is very broad and some
proposals appear paradoxical at first glance.

Feyman (1987) showed how *negative probabilities*—introduced since the
thirties within the context of quantum mechanics—naturally exist in the world and
presently researchers use negative values of probability to solve various issues
(Khrennikov 1999). *Comparative probabilities* (Fine 1973), *temporal probabilities*
(Haddawy 1994) and *extended probabilities* (Mückenheim 1986) emerged after
accurate theoretical investigations. The scenario turns out to be so wide and
complex that one risks of getting lost, but empirical observations in the field can
suggest a way out.

It may be said that the theories from 1 to 7 and even other theories on probability do not come into conflict with the mathematical calculus in a significant

manner. Mathematicians coming from various schools concur in the calculation of incompatible events, of conditional probabilities and other significant formulas. Von Mises, Popper, de Finetti and others tend to use uniform mathematical equations.

Debates arise on the nature of probability, and the so called *problem of interpretation* still appears to be unsettled.

The divergences existing among the schools 1–7 lead experts to recognize three major tendencies:

(i) *Objectives Theories* (see 1, 2 and 7) hold that probability is the objective property of a physical event.
(ii) *Formal Theories* (see 3 and 6) sustain that the probability is the measure of abstract sets or of linguistic systems.
(iii) *Subjective Theories* (see 4 and 5) interpret the concept of probability as the specification of a personal knowledge.

Pragmatic observations in the present statisticians' community show that constructions (ii) do not create great problems in that the logical theory seems declining while *de facto* Kolmogorov's axiomatization provides the mathematical base to several works. The essential aspect of the latter theory is that it is not necessary to give any philosophical rendering or comment on probability in relation to the world. One has to apply mathematical axioms and to perform appropriate operations of calculus.

Ultimately, constructions (i) and (iii) keep the problem of interpretation open so far. Frequentists see the probability as the limit value of the relative frequency in a *collective*. Subjectivists hold that the probability is the degree of a personal belief. The two schools of thought differ on which class of questions is more important to answer in scientific and engineering contexts, on the methods to follow, on the criteria to apply and other topics. The Bayesians and classical statistics which stay behind families (iii) and (i) respectively relaunch the debates of theorists. Several followers of schools (i) and (iii)—for the sake of simplicity, I shall overlook the differences between the Bayesian and the subjectivist, and amongst the various 'objective' theorists from now onward—do not miss any opportunity to underline the logical incompatibility of their own conceptualization with the opposite works.

The overall scenario appears somewhat rich and complex because, aside the experts who exacerbate the divisions, we find those who assume a somewhat 'ecumenical' behavior. It may be said that there are two scientific circles. The members of the first circle, whom I shall call *orthodox*, declare their fellowship to one school of thought; they sustain the validity of their distinguished position and reject the opponent's viewpoint as untrustworthy. The members of the second group—called *pluralists, dualists,* or even *eclectic* in the current literature—do not exclude the possibility of using various statistical formulations; in particular they tend to admit the frequentist and the subjective views alike on the basis of somewhat pragmatic criteria.

1.2 Pluralists

The circle of pluralists is second in the present order of illustration but not in the historical sense.

Pascal and other pioneers did not raise objections against the odd significance of the probability. They solved a variety of questions and became involved in the multifold notion of the probability as *chance*, as *odds,* as *expectation,* as *subjective judgment upon a wager*, as *calculus of combinatorics* and so forth (Hacking 1984).

During the nineteenth century several experts became more conscious of the multiple nature of probability and introduced expressions more close to the present terminology. Historians assign several writers to the community of pluralists, such as Jean–Antoine Condorcet, Joseph Louis Bertrand, Henry Poincaré, Antoine Cournot, Denis Poisson, Bernard Bolzano, Robert Leslie Ellis, Jacob Friedrich Fries and John Stuart Mill. These authors worked in France, Germany and Great Britain, independently one of the other and in various periods of time.

Even some inventors of theories 1–7 sustain the eclectic stance. Carnap introduces the terms probability$_1$ as *rational credibility* and probability$_2$ as limiting *relative frequency of occurrences*. Ramsey separates the probability used in experimental sciences related to *frequencies*, and the probability necessary to make decisions. Van den Hauwe (2007) tells that even von Mises embraced the pluralist philosophy in the last period of his life.

Modern thinkers who support the multifold concept of probability—as a principled rather than pragmatic view—meet non trivial obstacles. They attack the problem from various directions. Some try to place the two schools of thought closer together and encourage a 'philosophical compromise' such as Costantini (1982) who supports 'moderate' positions. Jeffreys (1955) sustains the idea of physical probabilities in between an epistemic framework. Lewis (1986)—starting from a subjectivist position—aims at showing that we really need two concepts of probability. Miller (1966) argues about the *probability coordination principle* and asserts one ought to set his subjective probability for an event equal to what he takes to be the physical probability of that event. A group of authors tend to bring closer the statistical schools on a technical plane; some introduce special techniques which could connect Bayesian and frequentist practice such as the bootstrap (Efron 2013) and the maximum entropy (Wolpert 1994); others investigate the fields where the two schools yield essentially the same answer such as chosen prior distributions (Cox 2005). A circle of thinkers remarks that different interpretations of probability are independent one of the other and can be applied in different contexts. Gillies (2000) holds that the subjective notion of probability is suitable for social sciences, whereas an objective notion fits in with natural sciences; all without committing any logical inconsistency.

The multifold, wide–ranging production of pluralist thinkers cannot easily be summed up. Salmon (1988) provides a tentative synthesis by subdividing the pluralist theories into *frequency–driven* (F–D) and *credence–driven* (C–D) ones. In F–D accounts, frequencies play a major role in determining probabilities, whereas

Table 1.1 Distribution of books

Year periods	Number of books			
	Pluralist	Bayesian	Frequentist	Total
1941–1966	0	1	19	20
1967–1980	6	2	26	34
1981–2008	173	37	188	398
	179 (0.39)	40 (0.08)	233 (0.51)	452

in C–D accounts probabilities are strongly related to belief. In reality, a summary of the works cannot be compiled at ease and I limit myself in presenting a bibliography in Appendix B. This partial and uncommented list intends to highlight the variety of positions and ideas devised by modern writers.

Eventually the circle of pluralists includes several pragmatic statisticians who tend to overlook theoretical discussions and to minimize the weight of the consequent debates. They often contend in private conversations that the multiple approaches to the probability interpretation turn out to be natural and rather obvious for experts who, in the past, saw the long dispute between E.S. Pearson and R.A. Fisher. This contrast came to an end when statisticians began to select the approach most appropriate to the solution of the problem in question. Pragmatic orientation enabled experts to overcome the past contrast, and some conclude that history will repeat itself.

A part of modern literature mirrors unconcern on the logical incongruence emerging among the various notions of probability. I conducted a special research on this issue and surveyed 452 books of statistics written for students and majors (Rocchi et al. 2010). The canvassed books are school-texts and technical manuals written by 645 authors who illustrate basic notions of statistics and advanced topics as well (Table. 1.1). The books published between 1941 and 2008 show an increasing tendency towards pluralism as time went by. The most recent sampled works present the mathematical weapons pertaining to both the schools, while the writers show blatant unconcern for the logical inconsistency emerging between the subjective and the frequentist theory. The authors are inclined to minimize the weight of the philosophical controversy extant at the base of the statistical methods in use and sometimes declare their personal opinion in explicit terms.

No pluralist theoretical proposal gained universal consensus so far. Reality shows that there is no easy settlement of the contrasting doctrines. Frequentists and subjectivists adopt diverging interpretations, theoretical constructions and distinct sets of mathematical tools. The answers to a number of queries are diametrically opposite. For instance:

"Is picking a lottery number that appeared less frequently over the last draws more favourable than a hot-number?" Frequentists say yes; Bayesians say no.

"Does there exist a true fixed and non-random population parameter, even if we cannot know its value because all we can see is the realizations of some random variable?" Frequentists say yes; Bayesians say no.

Due to the conflicting conclusions, pluralist statisticians should not use both the statistical methods as a sort of 'thinking caps' to put on in the place of each other. The two schools are really discussing different things and the usage of different statistics does not appear correct to me unless one explains why, when and how to adopt the pluralist philosophy. The compatibility of the frequentist and subjective viewpoints should be demonstrated throughout accurate studies, and the multiple views on probability should be justified in a rigorous epistemic sense.

I believe that a theoretical framework should sustain the pluralist mode and the first part of this book means to address this kind of problem.

References

Costantini, D., & Geymonat, L. (1982). *Filosofia della Probabilità*. Milano: Feltrinelli Editore.

Cox, D. R. (2005). Frequentist and bayesian statistics: A critique. In *Proceedings of the Statistical Problems in Particle Physics, Astrophysics and Cosmology* (pp. 3–6). London: Imperial College Press.

Eagle, A. (2004). Twenty-one arguments against propensity analyses of probability. *Erkenntnis, 60*(3), 371–416.

Edwards, P. (Ed.). (1967). *The encyclopedia of philosophy*. New Jersey: Prentice Hall.

Efron, B. (2013). A 250–year argument: Belief, behaviour and the bootstrap. *Bulletin of the American Mathematical Society, 50*(1), 129–146.

Feynman, R. P. (1987). Negative probability, in quantum implications. In F. D. Peat & B. Hiley (Eds.), *Quantum implications: Essays in honour of David Bohm* (pp. 235–248). London: Routledge & Kegan Paul Ltd.

Fine, T. L. (1973). *Theories of probability, an examination of foundations*. New York: Academic.

Gillies, D. (2000). Varieties of propensity. *British Journal for the Philosophy of Science, 51*, 807–835.

Hacking, I. (1984). *The emergence of probability: A philosophical study of early ideas about probability, induction and statistical inference*. Cambridge: Cambridge University Press.

Haddawy, P. (1994). *Representing plans under uncertainty: A logic of time, chance, and action*. Berlin: Springer.

Jeffreys, H. (1955). The present position in probability theory. *British Journal for the Philosophy of Science, 5*, 275–289.

Khrennikov, A. Y. (1999). *Interpretations of probability*. Tokyo: VSP International Science.

Lewis, D. (1986). A subjectivist's guide to objective chance. In D. Lewis (Ed.), *Philosophical papers* (Vol. 2, pp. 83–132). Oxford: Oxford University Press.

Maistrov, L. E. (1974). *Probability theory: A historical sketch*. New York: Academic.

Miller, D. (1966). A paradox of information. *British Journal for the Philosophy of Science, 17*(1), 59–61.

Mückenheim, W. (1986). A review of extended probabilities. *Physics Reports, 133*(6), 337–401.

Rocchi, P., Pandolfi, S., & Rocchi, L. (2010). Classical and bayesian statistics: A survey upon the pluralist production. *International Journal of Pure and Applied Mathematics, 58*(3), 255–280.

Salmon, W. C. (1988). Dynamic rationality. In J. Fetzer (Ed.), *Probability and causality*. Dordrecht: Reidel Publishing Company.

Van den Hauwe, L. (2007). John Maynard Keynes and Ludwig von Mises on probability. *MPRA Paper* No. 2271. Retrieved 2007 from http://mpra.ub.uni-muenchen.de/2271/.

Weatherford, R. (1982). *Philosophical foundations of probability theory*. London: Routledge and Kegan Paul Ltd.

Wolpert, D. (1994). Reconciling bayesian and non-bayesian analysis. In G. R. Heidbreder (Ed.), *Maximum entropy and bayesian methods* (pp. 79–86). Dordrecht: Kluwer.

Chapter 2
A Mathematical Approach to the Interpretation Problem

Before going into the proposal's content, I wish to give a brief account on the genesis of the present work.

2.1 Why I am a Pluralist

Experts apply classical statistics and Bayesian statistics in a myriad of situations and this apparent, global fact inspired some thoughts in me which I summarize as follows.

I saw the fortune of classical statistics and Bayesianism in the world as a massive test of the theoretical constructions that sustain statistical methods. It came to me: "If the radical philosophers were right—that is to say, if either the frequentist or the subjective theory is false—then several evidences capable of disproving either von Mises or de Finetti should have emerged from the numerous applications achieved by working statisticians". Instead, there is no definitive disproof against the former or the latter; scientists and professionals adopt both the statistical methods at ease in various circumstances.

Probability is central to many important sectors: economics, business, government, medicine, physics, engineering and many others (Gigerenzer 1989). We could say that statistics sustain the advancement of present-day economies and the global test achieved by modern statisticians has an incredibly huge dimension.

No logical frame sustains the pluralist thesis in a convincing manner, but I am persuaded that: "Opinions are disputable and facts are true" and facts show that the double approach to probability is working successfully. The pluralist statistical mode gets favorable feedback and—in my opinion—demonstrates that the Janus-faced nature of probability cannot be denied. I am inclined to attack the interpretation problem without any philosophical prejudice.

P. Rocchi, *Janus-Faced Probability*, DOI: 10.1007/978-3-319-04861-1_2,
© Springer International Publishing Switzerland 2014

2.2 Guidelines

Ramsey, Carnap and others accepted the pluralist hypothesis but in reality devoted themselves to a single side of the probability foundation; instead, Popper paid attention to the various aspects and did not exclude any subject to investigate. I felt that Popper was a sincere pluralist who methodically attempted to go toward the comprehensive conceptualization of probability, and I got significant help and guidance from his lesson.

I was working in the computer sector and shared the feeling of the Austrian thinker:

> In modern physics . . . we still lack a satisfactory, consistent definition of probability; or what amounts to much the same, we still lack a satisfactory axiomatic system for the calculus of probability; in consequence, physicists make much use of probabilities without being able to say, consistently, what they mean by 'probability' (Popper 2002).

What Popper wrote for physicists, I could repeat for computer engineers who were in need of clearer statistical notions.

When I became involved in the foundations of the probability calculus, Popper's writings provided a guide to my inquiries: The following points summarize my personal view on Popper's production and inspired precise conclusions to me:

(1) Popper distinguishes two kinds of topics when he focuses on the long–run event probability and on single event probability. Carnap deems this difference 'inessential' instead—in my opinion—*this subdivision suggests the analytical criterion to inquiry into the nature of probability*. On one side, I intend to consider repeated events and on the other single events.

(2) Popper (1959) defines probability as a physical propensity, or tendency of a given type of material situation to yield an outcome of a certain kind. From Popper's lesson I infer that *one should validate the calculated value of probability because of its physical origin*. In particular I share the idea that one should be able to control the probability value by means of tests conducted in the physical environment.

(3) For Popper (1954) the concept of probability is bound to the problem of knowledge and experimentation. He links the foundations of the probability calculus with the methodology of science, and I agree that *one should discuss the pluralist approach on probability in accordance with the praxis of science*.

The lessons of the Austrian philosopher gave orientation to my way of reasoning. The points listed above provided the seeds of my investigations. However, I do not recognize myself as an authentic follower of Popper in that I decided to adopt an autonomous method of study.

He develops his theory of propensity on the basis of philosophical considerations; instead, I have little confidence in the philosophical methods of inquiry and *resort to a purely analytical approach*. I mean to talk about the probability foundations through *two theorems* and a *property of functions*.

2.3 An Analytical Approach

I assume the Kolmogorov postulates are true in the present section of the book. I do not discuss the expressions in use such as the equations on conditional probability, the multiplication law, the stochastic systems, the probability functions and so on.

I mean to demonstrate two theorems on the basis of the following hypotheses:

• Let us examine the random event A such as

$$0 < P(A) < 1. \tag{2.1}$$

For the sake of simplicity, we suppose that the indeterminate events make a sequel of Bernouilli trials; that is, the experiment outcome can be either of two possible results: 'success' and 'failure'. By convention, the symbol A denotes the success and the relative frequency of success is

$$F(A_n) = N(A_n)/n. \tag{2.2}$$

where $N(A_n)$ is the number of success in a sequence of n trials. This sequence constitutes an application of independent and identically distributed (i.i.d.) random variables.

• In accordance with point (1) of the previous section, we assume the following mutually exclusive values. The trials occur several times

$$n \gg 1. \tag{2.3}$$

The single case event has only one trial

$$n = 1. \tag{2.4}$$

• As per point (2) above, we assume that a relationship exists between the probability and the relative frequency and we prove the following theorems dealing with this relationship.

2.3.1 Theorem of Large Numbers (Strong)

Suppose (2.3) true and consider the *Law of Large Numbers* (LLN) in the strong version demonstrated by Émile Borel (1909). This theorem asserts that the larger the number of repetitions of Bernoulli trials, the better the approximation of the relative frequency to the probability tends to be

$$P\left[\lim_{n \to \infty} F(A_n) = P(A)\right] = 1. \tag{2.5}$$

This is often abbreviated to

$$F(A_n) \overset{a.s.}{\to} P(A) \quad \text{as } n \to \infty \tag{2.6}$$

which means $F(A_n)$ converges almost surely to $P(A)$ as n tends to infinity.

Proof See Appendix C.

2.3.2 Theorem of a Single Number

Let (2.4) true, the *Theorem of a Single Numbers* states that the relative frequency is not equal to the probability

$$F(A_1) \neq P(A). \tag{2.7}$$

Proof The successful event A either occurs or does not occur when the operator arranges a sole trial. The following alternative values are allowed for the frequency

$$F(A_1) = \frac{N(A_1)}{1} = \frac{1}{1} = 1. \tag{2.8}$$

$$F(A_1) = \frac{N(A_1)}{1} = \frac{0}{1} = 0. \tag{2.9}$$

Inequality (2.1) excludes that A is sure or impossible, thus (2.8) and (2.9) mismatch with $P(A)$ that is decimal.

2.4 Two Remarks on Mathematics and Science

The mathematical language is a highly abstract symbol system. The above presented results appear extremely concise and are not so easy to argue over; hence I make a premise on the mathematical calculus and the scientific method which will give support to discussion.

This book matches with Popper's thought who saw probability in relation to science's procedures—see point (3). However, I do not intend to develop philosophical considerations, instead I mean to talk about a general property of functions.

2.4.1 Physical Mapping

By definition, a mathematical function (or transformation) establishes a directional relationship from the argument to the result (Royden and Fitzpatrick 2010). For the sake of simplicity, we take the following function

$$y = f(x). \tag{2.10}$$

Which maps the domain into the codomain

$$f : X \to Y.$$

Sometimes x is called *free* variable in opposition to y which is decided by x; notably, *the numerical value of y depends on the argument x, and also the mathematical typology of y relies on x.* If x is a scalar, then y is a scalar; if the argument is a vector, then the output is a vector and so forth.

Mapping turns out to be more complex in engineering, physics, biology etc. since a function fixes a double form of correspondence in those environments. *The argument x determines the mathematical value of the result y and its physical nature too.* As an example, take the relative variation of x respect to time

$$y = f(x, t) = \frac{\Delta x}{\Delta t}. \tag{2.11}$$

In abstract, we have the mathematical correspondence

$$f : X, T \to Y.$$

Professionals use (2.11) for an assortment of practical purposes and when x is the energy of a motor engine, one obtains the power W of that engine

$$W = \frac{\Delta E}{\Delta t}.$$

That is to say

$$f : E, T \to W.$$

This expression holds that the numeric value W relies on E and t, and also the material features of W depend on E and t. When the variable is the speed of a body, one gets the acceleration of that body

$$\vec{a} = \frac{\Delta \vec{v}}{\Delta t}.$$

The change of magnetic flux Φ produces the induced electromotive force ϵ

$$\epsilon = -\frac{\Delta \Phi}{\Delta t}.$$

The arguments E, v and Φ—besides time—determine the nature of W, a and ϵ respectively. The left side of an equation follows the right side, as well as regarding the mathematical and the material properties too. Two forms of obligatory dependence regulate a mathematical expression in applications:

(*a*) *Mathematical mapping*: the argument determines the mathematical properties of the result.
(*b*) *Physical mapping*: the argument fixes the concrete nature of the result.

Fig. 2.1 The parameters
Z,U,K providing further
illustration of Y

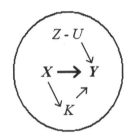

Property (*b*) appears evident to physicists who corroborate the dimensions of the equation members. Dimensional analysis is routinely used to check the plausibility of derived equations, on the basis that a result must have the dimensions of the argument. Actually, any quantity is the combination of the basic physical dimensions that form it, e.g. common elementary physical dimensions are length [L], mass [M], time [T], and electric charge [Q]. The input of an applied calculus has certain dimensions and the output has to have identical dimensions. As an example, one obtains the speed from the ratio distance/time and speed has dimension $[LT^{-1}]$. The dimensions of an argument can make a complex group such as $[T^{-1}M\,L^2Q^4]$ and in parallel the result has dimension $[T^{-1}M\,L^2Q^4]$.

Sometimes the argument settles a result that does not have any physical dimension but is only a pure number.

Finally the argument provides an outcome which is not workable or workable only under special conditions, and is called *imaginary*. As an example, the voltage in an alternating current (AC) circuit is given by a composite argument

$$V = V_0(\cos \omega t + j \sin \omega t).$$

The first part of the argument is real and denotes the regular component of the electrical current (Langsdorf 2001). The second part includes the imaginary number $j = \sqrt{-1}$ and calculates the component of the AC voltage that has the special property of being out of phase with the source. For this reason the imaginary term of V is sometimes thought as the "missing" voltage.

In conclusion, the argument of a function can generate an outcome:

(1) That has physical dimensions.
(2) That is dimensionless.
(3) That is imaginary.

Physical mapping can have a profound impact on the nature of the result since a theory usually comprises the relation $y = f(x)$ together with other equations which add further details to the correspondence $X \rightarrow Y$ (Fig. 2.1).

The argument x can modify the nature of the result y in a substantial manner because one or more physical laws, special properties or peculiar phenomena intervene in parallel to $y = f(x)$. The Boyle–Mariotte law offers a fine example (Ball 2002). This law states that the absolute pressure P and the volume V of an ideal gas are inversely proportional, when the temperature does not vary

Table 2.1 Two physical meanings of time

Assumption	Mathematical consequence	Physical consequence
$v \ll c$	$\Delta t_R \approx \Delta t_C$	*Classical time*
$v \approx c$	$\Delta t_R > \Delta t_C$	*Relativistic time*

$$V = \frac{k}{P} \qquad k > 0.$$

In addition experts discovered the *lines of equilibrium* between liquid/gas, liquid/solid, and solid/gas in the *phase diagram*. When P becomes very high—exactly when P surpasses the line of equilibrium in the phase space—the volume of gas changes its physical state and becomes liquid. In other words, when the pressure grows, it modifies the significance of the result V that ascertains the volume of a liquid instead of a sample of gas. The meaning of V changes in that the Boyle–Mariotte law and the phase diagram are simultaneously true.

I bring a second example. Let us work out the time Δt_R spent by a moving observer traveling at the speed v

$$\Delta t_R = \frac{\Delta t_C}{\sqrt{1 - \frac{v^2}{c^2}}}.$$

where c is the speed of light and Δt_C is the interval time when the observer is at rest. The more v gets close to the speed of light, the more time t_R dilates.

A second phenomenon occurs besides time dilation, the high speed v moves the observer from classical physical context to relativistic context (Wald 1984). Very high values of v result in astonishing phenomena where space is no longer a container and time is not an independent variable, but space and time make a *unicum* while gravity affects the shape of space and flow of time. Gravitational force bends the fabric of space and time, and the distortion of the space–time continuum even affects the behavior of light.

Relativistic universe is far dissimilar from classical universe. This just to say that the variable v affects the size of t_R and also its physical meaning; relativistic Δt_R differs significantly from classical Δt_C (Table 2.1).

The input to $y = f(x)$ affects or even subverts the nature of y because a theorem, or a physical law, or a special property etc. occurs in parallel to the variation in size of x. When the value of the free variable changes, the result varies numerically as stated in (*a*) and also can vary in qualitative terms as stated in (*b*).

Physical mapping applies to probability of course, and entails that *the physical meaning of probability does not exist by itself or in itself but depends on the probability argument*. In the light of the properties (*a*) and (*b*), the nature of probability does not tower as a metaphysical issue but proves to be a question which should be addressed on the basis of the probability argument.

2.4.2 Division of Labor

The property of mapping leads experts toward a precise division of jobs. Pure mathematicians study entirely abstract concepts, they have little interest in the physical nature of results and focus on abstract mapping (a). Experimentalists, engineers and professionals borrow the language from pure mathematicians, and in addition pay attention to how the results manifest in the real world and are deeply involved in both (a) and (b).

The two fields of study undertake different strategies. The abstract properties of a mathematical expression, which are indifferent to any material implication, cover a broad territory and are eternal. In contrast, the applied properties of a mathematical expression are narrower in scopes and change depending on the assigned argument. A single equation can convey several physical meanings in different contexts.

As an example, take the inversion (or negation) function usually written as NOT(a) in the Boolean Algebra where a is one or zero (Givant et al. 2010). Abstractly, the negation function inverts the binary value 1 to 0, and vice versa

$$\text{NOT } (0) = 1,$$
$$\text{NOT } (1) = 0.$$

This function has various general properties; perhaps the most famous one is the following

$$\text{NOT } (\text{NOT}(0)) = 0,$$
$$\text{NOT } (\text{NOT}(1)) = 1.$$

Philosophers popularize this rule by the famous aphorism: "A double negative says yes".

The inversion function can be used in the following applications:

- A sentence is true ($a = $ T) or false ($a = $ F) and its negation inverts the values of truth in classical logic.
- In linguistic, the binary variables stand for two adjacent words such as the subject ($a = $ S) and the verb ($a = $ V); and the inversion function interchanges the position of the words in a sequence. For example there are SV and VS linguistic structures in Italian.
- In digital technology, the bits are high–voltage value ($a = $ H) and low–voltage value ($a = $ L) and the gate NOT in a computer system swaps the bits over.
- In psychology, the binary value M or F denotes the sex, and its inversion depicts the taking on of the gender role of the opposite sex.
- In chemistry, the direction of optical rotation of a chemical substance may be dextrorotatory ($a = $ D) or levorotatory ($a = $ L). In chemistry, these terms refer, respectively, to the properties of rotating plane polarized light clockwise (for D) or counterclockwise (for L). NOT function reverses the optical rotation of a substance from D to L or vice versa.

The inverse function is unique in Boolean Algebra that treats NOT on the purely abstract plane; whereas NOT has several connotations in applications.

I could describe the purposes of an abstract mathematician as *global,* and the targets of an applied mathematician as *local.* All this entails a rather neat organization of labor which is flexible in the sense that an individual can pass from an abstract theorem to a concrete issue. He can change job and even can work in the two fields at the same time, anyway the duties which he fulfills, the methods of work and the objectives that he pursues in abstract mathematics and applications turn out to be rather diverging.

The comments presented in this section and in the preceding section will be used in the examination of the theorems of large numbers and a single number. These remarks will give some insights into the meaning of the theorems' conclusions.

2.5 Different Arguments, Different Probabilities

Theorems 2.3.1 and 2.3.2 establish the following relationships between the probability of a success and the relative frequency of that success

$$F(A_n) \xrightarrow{a.s.} P(A) \quad \text{as } n \to \infty, \tag{2.6}$$

$$F(A_1) \neq P(A). \tag{2.7}$$

These results have been proved through logical reasoning and are true beyond any doubt; they are fully accurate in mathematical terms but the same cannot be said from the physical viewpoint. Physical mapping *(b)* teaches us that the practical meaning of P should be established on the basis of each specific argument. Instead conclusions (2.6) and (2.7) refer $P(A)$ where the symbol A represents an abstract subset in the Kolmogorov theory. The probability $P(A)$ has no connection with a precise material situation. For instance, $P(A)$ may be tested once, one million times or even can never be subjected to any validation. The probability $P(A)$ does not say anything about the way it may be corroborated in the real world, instead the theorems assume that the proper objects of ascriptions of probability are series of experiments or single instances

$$n \gg 1, \tag{2.3}$$

$$n = 1. \tag{2.4}$$

Property *(b)* requires to specify that the theorem of large numbers applies exclusively to $P(A_n)$ and the single number theorem to $P(A_1)$; otherwise, the theorems risk providing generic and rather unintelligible accounts. Conclusions (2.6) and (2.7) are to be rewritten in the ensuing more accurate terms

$$F(A_n) \overset{a.s.}{\to} P(A_n) \quad \text{as } n \to \infty. \tag{2.12}$$

$$F(A_1) \neq P(A_1). \tag{2.13}$$

The first result demonstrates that *one can control the probability of repeated events by mean of observations undertaken in the physical reality*. The second result establishes that *never and ever one can corroborate the probability of a single experiment*. It is not a question of instruments or methods; it is simply impossible. The probabilities $P(A_n)$ and $P(A_1)$ have diverging behaviors where the difference is neat and one cannot assimilate $P(A_n)$ to $P(A_1)$, nor $P(A_1)$ to $P(A_n)$; hence, we have three kinds of probability in all.

By definition, the probability $P(A)$, which one calculates in abstract, is independent of a determined physical circumstance. Verbatim this matches with the adjective '*abstract*' which signifies something that has no precise connection with the world out there. Frequently authors compare the Kolmogorov theory to the pure geometry, both of them viewed as idealized models having only a loose relationship with reality. $P(A)$ is exclusive to the abstract formulation and is unique because it relies on univocal axioms. $P(A_n)$ and $P(A_1)$ are pertinent to applications.

Frequentists, subjectivists and the followers of Kolmogorov share the basic statements of probability such as the multiplication law, the Laplace formula etc. $P(A)$, $P(A_n)$ and $P(A_1)$ can be equal as numbers, but their substance is absolutely dissimilar; their practical features diverge in evident manner. $P(A_n)$ can be controlled by means of tests, whereas $P(A_1)$ cannot. By way of illustration, let us examine the probability of flipping a coin that comes up head (Dunn 2005). Since the possible cases are equiprobable we use the Laplace equation

$$P(A_H) = 1/2 = 0.5. \tag{2.14}$$

The relationship between $P(A_H)$ and a practical situation is any, by definition. Now we consider the probability of obtaining head after 100 coin flips.

Also $P(A_{H100})$ falls under the assumption of Laplace equation and is equal to 0.5; in addition $P(A_{H100})$ can be compared to $F(A_{H100})$ by means of experimental tests or using the Monte Carlo methods. Table 2.2 contains the real results obtained after ten series of one hundred trials. Where H is the number of heads which came up in each series, and D_H is the percentage difference from probability to the relative frequency of heads:

$$D_H = \frac{|P(A_{H100}) - F(A_{H100})| \cdot 100}{P(A_{H100})} = \frac{|0.5 - (H/100)| \cdot 100}{0.5} \%.$$

Table 2.2 exhibits 'small values' of D_H, and this negligible deviation matches with theorem 2.3.1. Once again, I emphasize how this phenomenon is alien to $P(A_H)$ which, by definition, says nothing about its controllability in the world.

Table 2.2 Ten series of a hundred trials

	R1	R2	R3	R4	R5	R6	R7	R8	R9	R10
H	46	55	51	43	49	51	43	52	49	49
D_H (%)	8	1	2	14	2	2	14	4	2	2
T	54	45	49	57	51	49	57	48	51	51

Table 2.3 Ten series of a single trial

	R11	R12	R13	R14	R15	R16	R17	R18	R19	R20
H	1	1	1	0	0	1	1	0	1	0
D_H (%)	100	100	100	100	100	100	100	100	100	100
T	0	0	0	1	1	0	0	1	0	1

Now, I calculate the probability of a single event using (2.14) and attempt to validate this equation by means of a sequel of tests. Table 2.3 exhibits the outcomes of ten experiments consisting of a single trial.

The percentage difference D_H ranges regularly 100 %.

$$D_H = \frac{|P(A_{H1}) - F(A_{H1})| \cdot 1}{P(A_{H1})} = \frac{|0.5 - (H/1)| \cdot 1}{0.5} = 100\,\%.$$

It is evident how $P(A_{H1})$ has not been validated by this practical experience in accordance with the theorem of a single number. Beyond any doubt the three forms of probability are the same as numbers but one can reasonably conclude:

- $P(A_H)$ lies apart from any concrete instance.
- $P(A_{H100})$ finds corroboration in the physical world.
- $P(A_{H1})$ is out of control.

Experts get identical values for $P(A)$, $P(A_n)$ and $P(A_1)$ when they use the classical equations of the probability calculus and even when they calculate equations pertaining to special sectors—for example—thermodynamics and quantum mechanics.

The difference amongst $P(A_n)$ and $P(A_1)$ sometimes appears astonishing from the physical point of view. For example, the standard Copenhagen interpretation holds that the normalized wavefunction of the single particle ς gives the probability amplitude for the position of that particle. Quantum theorists use the following probability density function of the position x of ς at the given time t_0

$$\rho(x) = |\psi(x, t_0)|^2.$$

Thus the probability that the particle ς is present inside the volume V at t_0 is obtained by this abstract calculus

$$P(V) = \int_V \rho(x)dx = \int_V |\psi(x, t_0)|^2 dx. \tag{2.15}$$

Fig. 2.2

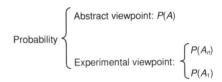

As a matter of fact, when an operator measures ç, it collapses. The particle ç assumes a certain position in the space and the statistical description (2.15) expires (Albert 2009). Various theories have been put forward to explain the specific cause of the particle's collapse; the specialist discussion focusing on quantum effects goes beyond the scopes of this book. What matters is that the collapse of the single particle is consistent with the present theory which holds that the probability distribution of a single particle is unreal. Experience shows that nobody can validate $P(V_1)$ using a single particle, instead several particles can confirm or disprove (2.15).

In conclusion, the reader sees three distinct parameters; and one cannot equate one to another even if sometimes they are identical in numerical terms. The probability $P(A)$ pertains to abstract investigations. It may be said that $P(A)$ is typical of *speculative mathematics*, whereas $P(A_n)$ and $P(A_1)$ conform to applied inquiries.

This kind of work subdivision is familiar to mathematicians. I mention the behavior of Newton and Liebniz as a comparative case in order to clarify what I am meaning to say.

In the 17th century there were heavy unsolved issues on mechanics. Experts were unable to calculate derivatives and integrals in a rigorous manner; they resorted to a set of visual and manual methods which provided approximate responses. Newton and Liebniz laid the basis for *arithmetica infinitorum* (= infinitesimal calculus) but they kept separate the abstract infinitesimal calculus from the applied calculus in mechanics. As an example, Newton fixed the *second law of mechanics*

$$F = \frac{d(mv)}{dt}.$$

But he did not mix up this result with the general notion of a derivative as a rate of change

$$y' = \frac{dy}{dx}.$$

Newton kept apart the applications of infinitesimal analysis from the general properties of infinitesimal analysis. In substance, this book means to introduce a similar division in probability and Fig. 2.2 separates the independent approaches that pertain to probability.

In the early 19th century German mathematicians pioneered this division of labor. A separation existed between theorists who proved theorems about probability and statisticians, professionals and scientists who employed the calculus in the world. Various experts belonging to the second group solicited a decision

Table 2.4 Two physical meanings of probability

Assumption	Mathematical consequence	Physical consequence
$n = 1$	$F(A_1) \neq P(A_1)$	*Untestable probability*
$n \to \infty$	$F(A_n) \to P(A_n)$	*Testable probability*

about the models of probability in order to apply the probability theory in the most appropriate manner.

More recently Doob emerges as the most active advocate of the division between real world probability and mathematical probability when the latter has the role of a measure theory of sets. Doob in (1996) comments on the developments put forward by Lebesgue, Borel, Radon, Fréchet, and the Radon–Nikodym theorem that paved the way to Kolmogorov axiomatization (see Appendix A). Doob also mentions the ongoing resistance by probabilists to an autonomous measure–theoretic framework.

Probability $P(A)$ associates a decimal number to the subset A and can be defined as a *two–termed functor* (Popper 1938)

$$f : A \to P. \tag{2.16}$$

where $A \in \Omega$, and $P \in \mathbb{R}$. The previous pages pinpoint that mathematical mapping regulates (2.16) in abstract calculations; mathematical and physical mapping govern $A \to P$ in applications. Physical mapping entails that the behavior of probability depends on the random event in a direct way. Chance cannot be tested on a single run and has a testable prediction on the long run; this special propensity of probability varies directly with the event size n. One can reasonably conclude that the increasing number of trials affects the nature of probability which evolves from being an untestable parameter to being a controlled one. This change—depending on n—can be compared to the nature of time which changes from classical to relativistic time when the speed rate v of the observer increases (Table 2.4).

Physical mapping leads us towards a position which differs from the perspective developed in philosophical treatises. Here the interpretations of probability do not emerge as absolute truths as orthodox philosophers hold, but are partial truths depending on the values of the involved parameters.

References

Albert, Z. D. (2009). *Quantum Mechanics and Experience*. Cambridge: Harward University Press.

Ball D. (2002). *Physical Chemistry*. California: Brooks/Cole.

Borel, E. (1909). Les Probabilités Dénombrables et Leurs Applications Arithmetique. *Rendiconti del Circolo Matematico di Palermo, 2*(27), 247–271.

Doob, J. (1996). The development of Rigor in mathematical probability (1900–1950). *The American Mathematical Monthly, 103*(7), 586–595.

Dunn, P. K. (2005). We can still learn about probability by rolling dice and tossing coins. *Teaching Statistics, 27*, 37–41.

Gigerenzer G. (1989). *The Empire of Chance: How Probability Changed Science and Everyday Life*. Cambridge: Cambridge University Press.

Givant S., Halmos P. (2010). *Introduction to Boolean Algebras*. New York: Springer.

Langsdorf A. S. (2001). *Theory of Alternating Current Machinery*. New York: McGraw–Hill.

Popper, K. R. (1938). A set of independent axioms for probability. *Mind, 47*(186), 275–277.

Popper, K. R. (1954). Degree of confirmation. *The British Journal for the Philosophy of Science, 5*(18), 143–149.

Popper, K. R. (1959). The propensity interpretation of probability. *The British Journal for the Philosophy of Science, 10*(37), 25–42.

Popper K. R. (2002). *The Logic of Scientific Discovery*. London: Routledge.

Royden H., Fitzpatrick P. (2010). *Real Analysis,* Pearson (4th ed.). Upper Saddle River: Pearson.

Wald R. M. (1984). *General Relativity*. Chicago: University of Chicago Press.

Chapter 3
Probability Validation

The theorem of large numbers and the theorem of a single number that illustrate the possibility of determining the presence of probability in the physical reality are worthy of further examination.

Popper relates the concept of probability to the methodology of science, and a comment on the scientific praxis should be used to introduce the discussion of the theorems in an appropriate manner.

3.1 A Parameter that Cannot be Validated Does Not Exist

The Oxford English Dictionary (2012) says that scientific method consists in:

> Systematic observation, measurement, and experiment, and the formulation, testing, and modification of hypotheses.

Experimentation is the beginning and the end, the alpha and the omega for scientists who discover a phenomenon and verify it. When practical results contradict predictions, the hypotheses are called into question and explanations are to be sought.

Testing is supreme in sciences, and when one cannot corroborate the relation $y = f(x)$ because the measurements mismatch with the theoretical definition and there is no other plausible justification, than the theory is no longer valid and researchers conclude that *y is not extant in the world out there.* Popper underlines that only one test is sufficient to disprove a theory.

I give a pair of cases to clarify this procedure:

- Young and Fresnel replaced Newton's light corpuscles by waves propagating through flimsy material substance named *ether* (Newburgh 1974). They presumed that the propagation of light required a physical medium that was in a state of absolute rest and the ether concept became especially predominant by the end of the 19th century.

P. Rocchi, *Janus-Faced Probability*, DOI: 10.1007/978-3-319-04861-1_3,
© Springer International Publishing Switzerland 2014

However no evidence confirming the presence of an intervening substance through which signals can travel was found and so the concept of ether was abandoned. Today no-one mentions the ether whereas this idea was very popular in the past.

- In the twenties, the psychologist Hermann Swoboda and the biologist Wilhelm Fliess independently from each other, started investigations on the cycles that affect the internal functioning of the body and human behavior, particularly the physical, emotional and intellectual mental abilities. Those researchers traced *the theory of biorhythms*. The scientific community and even ordinary people paid a certain attention to the cycles of life but researchers did not find supporting evidence for this theory (Laxenaire et al. 1983). Several tests have been developed for detection and measurement of biorhythms but no result can sustain this notion, and in consequence biorhythms do not exist from the scientific viewpoint.

Summing up, scientists recognize the quantity y calculated by the function $y = f(x)$ as a real entity only if y can be measured in the world and if an assortment of tests can validate it. Under different circumstances, scientists reject y as a nonexistent entity.

This criterion will be used to examine furthermore the significance of the Theorems 2.3.1 and 2.3.2 and will cast new light on the conclusions just drawn.

3.2 Theoretical Proof and Empirical Validation

It may be said that the law of large numbers (LLN) pertains to the history of mankind. Since time immemorial, people observed the stabilization of the relative frequency in a sequel of identical trials.

The earliest form of this theorem was proved by Jacob Bernoulli who spent over 20 years to develop a sufficiently rigorous mathematical proof. His work was published posthumously in 'Ars Conjectandi' with a foreword by his nephew in 1713. This work impacted many mathematicians, in particular the 20th century saw numerous advances in probability which can be traced back to the Bernoulli theorem in some way (Denker 2013).

Various eminent authors contributed to improve the demonstration of LLN. In 1866 Lvovich Chebychev discovered the method which now bears his name. Later on one of his students, Andrey Markov observed that Chebychev's reasoning can be used to extend the theorem to dependent random variables as well.

In 1909, the variant of LLN—cited in Sect. 2.3.1—was formulated and proved by É. Borel in the context of the Bernoulli scheme. In 1926 Kolmogorov derived conditions that were necessary and sufficient for a set of mutually independent random variables to obey the law. Kolmogorov established the convergence of the sequence $\Sigma \sigma_k^2 / k^2$—sometimes called the *Kolmogorov criterion*—which is a sufficient condition for the law of large numbers to be applied to the sequence of mutually independent random variables X_k with variances σ_k (Pollett 1989).

A few years later, Aleksandr Y. Khintchine demonstrated a weaker version of LLN and established the division between the two prominent forms of the theorem called *weak* and *strong* respectively. The strong form asserts that the sample average converges almost surely to the expected value μ

$$\bar{X}_n = \frac{\sum\limits_{j}^{n} X_j}{n},$$

$$P\left(\lim_{n \to \infty} \bar{X}_n = \mu\right) = 1.$$

The sample average tends to μ with probability one. The weak law states that sample average converges in probability towards the expected value

$$\lim_{n \to \infty} P(|\bar{X}_n - \mu| > \varepsilon) = 0.$$

This result does not say that the average $\bar{X}_n$ is bound to stay near to the expected value if the number of trials increases. The weak theorem is satisfied for a given ε in a certain number of trials n; thus, if additional trials are conducted up to the number $(n + m)$, the weak law does not guarantee $\bar{X}_{n+m}$ that the new average is bound to stay near μ for such trials.

The strong and weak forms do not describe different laws but refer to different ways of describing the mode of convergence of the cumulative sample means to the expected value, and the strong form implies the weak one.

Since time before recorded history, people sensed that if an experiment is repeated a large number of times then the relative frequency with which an event occurs tends toward the theoretical possibilities of that event (Stigler 1990). When the probability calculus became an official sector of mathematics, researchers began to verify LLN more closely. Commentators mention the classical experiments undertaken by the French Georges–Louis Leclerc de Buffon, the English Karl Pearson, and by the South African John Kerrich who performed 4,040 coin tosses, 24,000 tosses and 10,000 tosses respectively (Epstein 2009). The English biologist W.F.R. Weldon recorded 26,306 throws of 12 dice, and the Swiss scientist Rudolf Wolf recorded 100,000 manual throws of a single die.

These classical experiments and many others show how LLN yields reliable forecasts and predictions.

3.3 From Controllability to Realism

The reader could question:
What does exactly follow from the theorem of large numbers?

The preceding remarks on the scientific method allow me to create forward chained inferences as follows:

- Test is supreme in science. Any theoretical quantity must be approved or disproved using experiments and Theorem 2.3.1 entails that—at least in principle—probability can be confirmed or falsified by means of practical observations. The relative frequency obtained after several repeated trials in a Bernoulli series bring evidence sufficient to accept or refute $P(A_n)$. We have in hand the practical system to validate $P(A_n)$ in the real world and this conforms to the scientific method in a perfect manner.
- From this substantial experiment, one can conclude that $P(A_n)$—only $P(A_n)$ and not any generic probability—is an authentic measurable quantity that exists independently of any observer when n is 'very large'.

The probability of a long term event is a real quantity in the world.   (3.1)

The theorem of large numbers shows that $P(A_n)$ is a physical parameter through thorough demonstration process, hence this result has the rank of universal truth in the literature. Writers unanimously follow Poisson (1837) who first equalized the Bernoulli theorem to a law of Nature:

Les choses de toutes natures sont soumises à une loi universelle qu'on peut appeler la loi des grands nombres (All kinds of things are subject to a universal law that can be called the law of large numbers).

- Statement (3.1) entails that one can deduce the probability from experiments. Actual experimentation furnishes the *empirical probability* $F(A_n)$ and the larger the number of trials, the more accurately one expects $F(A_n)$ to approximate $P(A)$. The relative frequency provides a reasonable value of the probability.

3.3.1 Statistics of Long-Term Events

Probability is the mathematical foundation upon which statistics depends. It may be said that probability is the heart and soul of statistics.

Statistical data are frequently analyzed to see whether conclusions can be drawn legitimately about a particular phenomenon and also to make predictions about future events. Clear probability principles help one to realize what an analysis really means and, sometimes even more important, what it does not mean.

Classical statistics—originated in the works of Pearson, Neyman, Fisher and others—operates under the assumption that an experiment is at least in principle, repeatable. This approach could be judged as an inferential approach which employs sampled data as its relevant source of information.

Due to the central position of the repeated sampling principle, experts assess the sample size together with other factors that influence sampling such as:

- Cost considerations (e.g. budget, desire to minimize cost).
- Administrative concerns (e.g. complexity of the design, research deadlines).
- Minimum acceptable level of precision.
- Confidence level.
- Variability within the population of interest.
- Sampling method.

These factors interact in complex ways and statisticians divide the methods for test which treat *ample samples* apart from those which treat *small samples*. The notions of ample/small samples derive from the central limit theorem. Normally a sample including 30 elements is deemed 'large enough' for the underlying distribution to be well approximated by the Gaussian one. In case of small samples, it is not possible to assume that the random sampling distribution of a statistics is normal and that the sampled values are sufficiently close to population values.

Increasing sample size is often the easiest way to boost the statistical power of a test. Higher sample size allows the researcher to improve the significance level of the findings; the test of hypothesis becomes more sensitive; and so confidence of the result becomes better (Ziliak and McCloskey 2008).

The central limit theorem is a cornerstone of classical statistics which includes interval and point estimators, tests of significance and tests of hypothesis. The optimal inference procedure is to be identified before the observations of data are available. One invents estimators of population parameters which conform to certain 'desirable' criteria like consistency, efficiency and lack of bias, and there is no systematic procedure of constructing estimators. Computer programs provide tools such as bootstraps and jack-knives. This just to say that statisticians have a large and rather loose collection of techniques at disposal. Table 3.1 exhibits a concise overview of the test methods available in traditional statistics, and gives an idea of the various procedures in use to a non-specialist reader.

In the present logical frame, it is significant that classical statistics is based on the assumption of infinite samples and in evident manner is consistent with the theorem that Bernoulli proved in 'Ars Conjectandi'.

3.4 From Uncontrollability to Unrealism

The experimental confirmation or denial of theoretical results is an essential passage in science. This principle even casts light on the meaning of the theorem of a single number which leads to somewhat astonishing and counterintuitive conclusions:

- The theorem of a single number says that one cannot validate the probability $P(A_1)$ in the world as the frequency $F(A_1)$ systematically mismatches with the

Table 3.1 Selecting a statistical test (from Motulsky 2010)

Goal	Type of data		
	Measurement from Gaussian population	Rank, score, or measurement from non-Gaussian population	Binomial
Describe one group	Mean, SD	Median, interquartile range	Proportion
Compare one group to a hypothetical value	One-sample t test	Wilcoxon test	Chi-square or Binomial test
Compare two unpaired groups	Unpaired t test	Mann-Whitney test	Fisher's test (chi-square for large samples)
Compare two paired groups	Paired t test	Wilcoxon test	McNemar's test
Compare three or more unmatched groups	One-way ANOVA	Kruskal-Wallis test	Chi-square test
Compare three or more matched groups	Repeated-measures ANOVA	Friedman test	Cochrane Q
Quantify association between two variables	Pearson correlation	Spearman correlation	Contingency coefficients
Predict value from another measured variable	Simple linear regression or Non-linear regression	Nonparametric regression	Simple logistic regression
Predict value from several measured or binomial variables	Multiple linear regression or Multiple non-linear regression		Multiple logistic regression

probability of A_1. It is not a question of inaccuracy; neither is it a question of operational obstacles, errors of instruments, or other forms of occasional limitation. Due to the essential role played by experimental validation in science, one necessarily reaches the following conclusion

The probability of a single event does not exist in the world as a physical quantity.

(3.2)

One can object that even frequentist probabilities may not show up in n trials. For instance, the probability of 1/6 for a certain number on a throw of a fair dice cannot show up in 10.000 trials because 10.000 is not divisible by 6.
The answer to this objection is the following.
The inequality $F(A_1) \neq P(A_1)$establishes a general rule, whereas the above mentioned difference between $P(A_n)$ and $F(A_n)$ depends on special values of n and cannot be proved in general. Hence, conclusion (3.2) applies to $P(A_1)$ and cannot regard $P(A_n)$.

• Statement (3.2) turns out to be dramatic since people often focus on a specific event and wish to forecast the outcome of this unique occasion. Sometimes the single event runs the risk of being essential for the survival of an individual or a business. Thus, there is a widely diffused desire to circumvent statement (3.2) and this desire has found support in the researches of subjectivist thinkers. The bypass devised by subjectivists can be illustrated in the following manner.

The value $P(A_1)$ does not exist as a physical quantity but is usually written down and as such it works as a piece of information

The probability of a single event can be used in communication. (3.3)

• As the probability of a single experiment is a linguistic unit, it carries information, it has a communication value and can express a personal belief. It is evident that people who assess $P(A_1)$—despite its uncontrollability in the physical world—trust in the number that they have written down or communicated vocally. Otherwise, $P(A_1)$ should become just pure nonsense. The symbol $P(A_1)$ manifests the credence of an individual thus it has a personal value and subjectivist philosophers conclude

The probability of a single event is subjective. (3.4)

Subjective probability has several significant antecedents. Besides generic and rather implicit notions which emerge in the production of Pascal and Laplace, we find explicit statements in the literature from the 19th century onward. Poisson (1837) writes:

The probability of an event is the reason we have to believe that it has taken place, or that it will take place.

Subjective probability could vary from person to person; but subjectivists do not deem 'the degree of belief' as something measured by strength of feeling, but in terms of rational and coherent behaviour (see Appendix A). Ramsey, de Finetti and Bayesians reply to the accusation of arbitrariness by introducing precise criteria to evaluate $P(A_1)$. The *Dutch Book criterion*, the *utility criterion* and the *plausibility of a belief,* according to past experiences, sustain the subjective modes.

Subjectivists gave significant insights into some aspects of the probability theory.

They corrected the 'law of averages'—a rather erroneous idea derived from the frequentist conceptualization—introducing the famous aphorism: "Probability has no memory". Previous trials cannot influence the next trial whose $P(A)$ cannot vary.

De Finetti (1930) perfected this notion by means of the *principle of exchangeability*; that is, a sequence of random outcomes is a sequence where future samples behave like earlier samples, and formally means that any order of a finite number of samples is equally likely (Kallenberg 2005).

3.4.1 Statistics of Single Events

Bayesians see $P(A_1)$ as a piece of information (see (3.3)); it follows by necessity that the statistical methods referring to A_1 center on the acquisition of information and on updating information after the test has been done. The statistics inaugurated by Savage is based on past and future observations, rather than on models, since exchangeable sequences cannot be modeled as samples from a fixed population.

The key to information acquisition is the specification of the *prior probability distribution* on θ, before the data analysis. There are various criteria to specify the prior distribution which can be done in some objective or subjective ways, and reflect the statistician's own prior state of belief. The Bayesian inference is the formalization of how the prior distribution changes to the *posterior distribution*, in the light of the evidence presented by the available data x as per Bayes' formula. In brief, the principal elements of Bayesian inference could be stated as follow.

There is an unknown state of nature which one wishes to learn more about:

- The agent specifies a probability model for unknown parameter values that includes some prior knowledge about the parameters if available. Prior beliefs about θ can be expressed as a probability distribution $\pi(\theta)$.
- The agent updates knowledge about the unknown parameters by conditioning the probability model $\pi(\theta)$ on observed data. Every observed x tells him something about θ.
- The agent updates his prior belief $\pi(\theta)$ to posterior belief $\pi(\theta|x)$.

In the Bayesian approach one considers the population parameters as random variables and studies their distributions in the light of the data. One infers the posterior distribution over quantities of interest; in detail, one evaluates the fit of the model to the data and the sensitivity of the conclusions to the assumptions. Complete enumeration of the posterior distribution in the entire hypothesis space is a computationally difficult task. This leads Bayesians to approximate methods like maximum likelihood which has inherent flaws and should be used carefully.

Once the assumptions about the model and the data are made, inferences are automatic. Whilst classical statisticians employ some inconsistent kinds of techniques, Bayesians enjoy a coherent, self–contained system. For frequentists, parameters are fixed and remain constant during this repeatable process; for Bayesians, parameters are unknown and are described probabilistically; data are fixed (Sheskin 2007).

A Bayesian gains information from previous experiments and thus he cannot calculate an event that has no equivalent. Burdzy (2009) puts forward a provocative remark: the subjectivist approach has the same problem as von Mises' theory; it does not provide probability for a unique event.

For the scopes of the present work, it is noticeable that the theorem of single number and the concept of $P(A_1)$ as a piece of information provide the fundamental guidelines to the logic of Bayesian statistics.

Concluding, the logic of experimental science spells out that the theorems convey two very profound and astonishing results: $P(A_n)$ is a real measure, $P(A_1)$ is a quantity which 'does not exist' in the world, hence one cannot parallel or put close $P(A_n)$ and $P(A_1)$.

By way of illustration, I go back to the probability $P(A_{H100})$ of obtaining head after 100 coin flips and the probability $P(A_{H1})$ of obtaining head in a single event just discussed in Sect. 2.5. Some subjectivist believes that $P(A_{H100})$ is the same as the repetition of $P(A_{H1})$ 100 times. Instead this is an absurd: *one hundred values that are out of control cannot make a single controllable measure.* One hundred values that 'do not exist' unfit a real and observable parameter. One cannot explain $P(A_n)$ using $P(A_1)$, neither one can refer the latter to the former because the natures of $P(A_n)$ and $P(A_1)$ lack features that can be equated and are suitable for comparison.

3.5 Sorry Science

The frequentist theory and classical statistics on the one hand and the subjectivist theory and Bayesian statistics on the other hand, appear similar to two distinct pillars that pertain to the same field of study which Costantini (1982) calls '*Science of Indeterminism*'. Just as many pillars sustain an ancient Greek temple, two huge columns should support the science of indeterminism. We could finally go to create a science that puts the two parts together. I personally believe that the entire ensemble of mathematical tools and methods could become more effective than they are now, if we were able to define a unified discipline.

However there are still some obstacles to overcome along the way.

Fig. 3.1

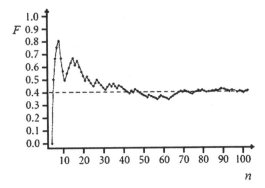

3.5.1 Severe Constraint

Statement (3.1) does not provide conclusions for any kind of events, but only for repeated events. This constraint was perfectly understood by von Mises who restricted his definition of probability to the collective; he even thwarted Fisher statistics on small samples because they do not match with the theorem of large numbers.

The starting point of the theorem of large numbers is the following assumption

$$n \gg 1. \tag{3.5}$$

This constitutes a rather severe limitation which becomes worse when limit (2.5) is true. The LLN is applicable under the following constraint

$$n \to \infty. \tag{3.6}$$

The perfect control of the probability occurs when the number of random events is infinite and this boundary condition can never be achieved. Countless trials should corroborate the quantity $P(A_n)$ but no scientist is able to achieve this ideal experiment.

A long-term event consists of a sequel of occurrences that, by definition, do not have any regular behavior and consequently $F(A_n)$ does not get close to $P(A_n)$ by following a regular approach. The frequency fluctuates as long as n becomes greater. Figure 3.1 exhibits a school case for sets of trials which vary from one to one hundred.

The validation of probability makes life hard for scientists and professionals, and in the past it created a lack of confidence in the science of indeterminism as an authentic science. See the criticisms raised by Von Kries (2009) by the end of the 19th century.

How can one disprove a probabilistic statement if tests vary irregularly and should never end?

Experts circumvent restraint (3.6) by means of progressive criteria. The larger n, the more the probability $P(A_n)$ finds confirmation and one selects the adequate degree of approximation to validate or disprove the probability in the world.

Statisticians suggest appropriate likelihood criteria; however, laymen find unmanageable those methods in everyday life. The distance between the expectation of people and the capabilities of the indeterministic mode cannot be easily filled.

3.5.2 Opponents to Unrealism

De Finetti's theory is openly grounded on the 'scandalous' conviction that there are no correct and rational probability assignments:

> The subjective theory (...) does not contend that the opinions about probability are uniquely determined and justifiable. Probability does not correspond to a self-proclaimed 'rational' belief, but to the effective personal belief of anyone. (de Finetti 1951)

Recently Galavotti (1989) furnished a fine comment on de Finetti's philosophy which

> Can be qualified as a combination of empiricism and pragmatism within an entirely coherent anti-realistic perspective.

The unrealism of single-event previsions turns out to be a limitation even more severe to the science of indeterminism than the limit condition $n \rightarrow \infty$ previously commented on. The incompatibility of indeterminate reasoning with reality was so impressive that philosophers rejected any form indeterministic logic and sustained the universal use of apodictic logic. For centuries, human though was dominated by determinism.

Determinism is the general philosophical thesis which states that for anything which ever happens there are conditions such that, given them, nothing else could happen (Earman 1986). There have been many versions of deterministic theories in the history of philosophy, springing from diverse motives and considerations, some of which overlap considerably. Different topics are drawn to the attention of commentators. Various questions upon free will, the Heisenberg principle, fatalism and other topics are still open and give substance to the ethical determinism, the logical determinism, the theological determinism, the physical determinism and other specific veins of study.

There are various degrees of adhesion to determinism inside this ample scenario. William (1956) recognizes the *hard determinists* who defend the strongest version of determinism. This circle of thinkers holds that every event has an antecedent cause in an infinite causal chain. Hard theorists assume that all propositions, whether about the past, present, or future, are either true or false. They sometimes reject free will and even refuse ethics. By *soft determinism* William James means all those theorists who hold that determinism is true and then by means of some elements man keeps a semblance of certain moral notions like liberty, responsibility etc. Scientists who are soft advocates of determinism alternate between chance and necessity, where the first results in unpredictable changes and the second maintains the world within rigid rules.

Determinism is something more influential than a rich, enduring school of thought. Determinism affected the use of reason, especially to form conclusions, inferences, and judgments in the entire Western philosophy. Nearly all ancient thinkers shared deterministic logic in a way that philosophers rejected any form of uncertain topics for centuries. They kept apart fuzzy thinking and elected apodictic reasoning as the systematic way to truth. Premises and conclusions had to appear unequivocal in order to avoid misleading. The apodictic style dominated the philosophical discussion everywhere and kept thinkers from paying sufficient attention to undecided situations (Loewer 2004). Libertarians and moralists, empiricists, logicians and many others believed that arguing about uncertain assumptions could yield not only generic conclusions but even raw errors. The following case—shared by von Mises—provided undeniable evidence for this opinion.

Suppose an urn includes a white ball and a black ball, and one forecasts two equally possible outcomes on the theoretical plane. Two results are possible as the evident essence of indeterminist mechanism. The black ball will potentially be drawn and at the same the white ball could be drawn. Despite the description of two equivalent states, only one result becomes true. Facts exhibit an outcome far differing from previsions. The black and the white balls do not generate a balanced outcome, but only one of them will come out of the urn.

Similar cases disproved the realism of indeterministic reasoning in the eyes of philosophers. They discovered that inferences upon single random occurrences led to an incompatible conclusion with reality. Broad considerations based on a state of incomplete knowledge yield conclusions which can mismatch with experience. More recent contribution on indeterminism and antirealism can be found in the essays by Donald Davidson (2001).

3.5.3 Things are No Longer the Same

The science of indeterminism could have potentially come to light for centuries but the severe constraints required to test $P(A_n)$ and the unrealism of $P(A_1)$, barred the progress of the studies. These logical and practical limitations prevented and indeed still prevent the science of indeterminism from maturing as a unified scientific discipline. We have seen how the vast majority of philosophers rejected the indeterministic reasoning because they considered it unreal to treat. An old saying tells: "Nature abhors a vacuum" and in a similar manner one could say that past philosophers 'abhorred indeterminism' from the period of classical Greek philosophy up to more recent times. Mathematicians paid little attention to the problems of chance until Pascal's time. Hacking (1984) annotates that during the Renaissance the calculus of probability was a poor science similar to alchemy, and much more untrustworthy than mechanics and astronomy.

Things did not change much until physicists and later economists, sociologists and other wide groups of professionals were induced to adopt statistics as substantial support to their works. A significant turning point arrived in the late 19th century. However, the mounting popularity did not ameliorate the scarce reputation in the probability sector which still remained in the air. Venn (1886) wrote:

> Probability has been very much abandoned to mathematicians, who as mathematicians have generally been unwilling to treat it thoroughly. They have worked out its results, it is true, with wonderful acuteness, and the greatest ingenuity has been shown in solving various problems that arose, and deducing subordinate rules. And this was all that they could in fairness be expected to do. Any subject which has been discussed by such men as Laplace and Poisson, and on which they have exhausted all their power of analysis, could not fail to be profoundly treated, so far as it fell within their province. But from this province the real principles of science have generally been excluded, or so meagrely discussed that they had better have been omitted altogether.

Nowadays things are no longer the same. Statistics and probability calculus have the distinction of being by far the most widely-applied mathematics; statisticians anticipate likely outcomes in virtually every field, from accounting to zoology, and from nuclear physics to medicine. They provide the necessary tools to understand and manage modern economies. Thinkers accept the indeterministic logic and researchers even devised novel intriguing theories. I confine myself to mention the *fuzzy logic* adopted in various advanced fields such as Artificial Intelligence and software programming (Kosko 1993).

We can guess that this astonishing progress will pave the way toward the unified science of indeterminism.

3.6 Closing Remarks

How to interpret statements of probability and what the proper objects of ascriptions of probability are, appear distinct topics in themselves and thinkers tackle them as independent issues so far. Instead, the present account deviates from this habit and has shown how the former follows the latter. It has been illustrated how the various meanings of probability logically derives from the features of random events. In particular, the concrete nature of $P(A_n)$ depends on the controllability of a large number n of trials. The subjective nature of probability comes from the unachievable control of $P(A_1)$, and in turn from the assumption $n = 1$.

This means that frequentism and subjectivism do not constitute absolute truths as several authors are inclined to believe so far, but rely on distinct mathematical hypotheses.

As a consequence of the autonomous hypotheses, it is natural to adopt two statistical methodologies which impose far different behaviours to practitioners. Classical statisticians believe that nothing is more important than repeatability, no matter what we pay for it; data come from a repeatable random sample and without any information being available prior to the model specification. Bayesians assume

Table 3.2 Separate inferences from different assumptions

1st step	$n \gg 1$	$n = 1$
	⇓	⇓
2nd step	$F(A_n) \rightarrow P(A_n)$	$F(A_1) \neq P(A_1)$
	⇓	⇓
3rd step	$P(A_n)$ can be controlled	$P(A_1)$ is out of control
	⇓	⇓
4th step	$P(A_n)$ is a real quantity	$P(A_1)$ is not a real quantity
		⇓
5th step		$P(A_1)$ is subjective

that prior information generally abounds and it is important and efficacious to use it; prior distribution is then updated with the obtained data (Table 3.2).

Concluding, two logical pathways, which match with universal experience, demonstrate to be different because the theoretical principles on which they are based are different.

References

Burdzy, K. (2009). *The search for certainty: On the clash of science and philosophy of probability*. Singapore: World Scientific.

Costantini, D., & Geymonat, L. (1982). *Filosofia della Probabilità*. Feltrinelli Editore.

Davidson, D. (2001). Indeterminism and antirealism. *Subjective, intersubjective, objective, philosophical essays* (Vol. 3). Oxford: Oxford University Press.

De Finetti, B. (1930). Problemi Determinati e Indeterminati nel Calcolo delle Probabilità. *Rendiconti Reale Accad. Naz. dei Lincei*, Ser. 6(12), 367–373.

De Finetti, B. (1951). Recent suggestions for the reconciliation of theories of probability. In *Proceedings of the Second Berkeley Symposium on Mathematical Statistics and Probability* (pp. 217–225). Berkeley: University of California Press.

Denker, M. (2013). Tercentennial anniversary of Bernoulli's law of large numbers. *Bulletin of the American Mathematical Society, 50*, 373–390.

Earman, J. (1986). *A primer on determinism*. Dordrecht: Reidel.

Epstein, R. A. (2009). *The theory of gambling and statistical logic*. Amsterdam: Elsevier.

Galavotti, M. C. (1989). Anti-realism in the philosophy of probability: Bruno de Finetti's Subjectivism. *Erkenntnis, 31*(2–3), 239–261.

Kallenberg, O. (2005). *Probabilistic symmetries and invariance principles*. New York: Springer.

Kosko, B. (1993). *Fuzzy thinking*. California: Hyperion.

Hacking, I. (1984). *The emergence of probability: A philosophical study of early ideas about probability, induction and statistical inference*. Cambridge: Cambridge University Press.

Laxenaire, M., & Laurent, O. (1983). What is the current thinking on the biorhythm theory? *Annales Medico-Psychologiques (Paris), 141*(4), 425–429.

Loewer, B. (2004). Determinism and chance. *Studies in History and Philosophy of Modern Physics, 32*, 609–620.

Motulsky, H. J. (2010). *Intuitive Biostatistic: A Nonmathematical Guide to Statistical Thinking* (2nd ed.). Oxford: Oxford University Press.

Newburgh, R. (1974). Fresnel drag and the principle of relativity. *Isis, 65*(3), 379–386.

Oxford English Dictionary (2012). Oxford University Press; available at: http://oxforddictionaries.com.

Poisson, S. D. (1837). *Recherches sur la Probabilité des Jugements en Matière Criminelle et en Matière Civile.* Bachelier.

Pollett, P. K. (1989). The generalized Kolmogorov criterion. *Stochastic Processes and their Applications, 33*(1), 29–44.

Sheskin, D. J. (2007). *Handbook of parametric and nonparametric statistical procedures.* Boca Raton: CRC.

Stigler, S. M. (1990). *The history of statistics: The measurement of uncertainty before 1900.* Cambridge: Harward University Press.

Venn, J. (1886). *The logic of chance.* London: Macmillan Company.

Von Kries, J. (2009). *Die Principien Der Wahrscheinlichkeitsrechnung: Eine Logische Untersuchung.* Charleston: Bibliobazaar Publisher.

William, J. (1956). *The will to believe and other essays in popular philosophy, and human immortality.* New York: Courier Dover Publications.

Ziliak, S. T., & McCloskey, D. N. (2008). *The cult of statistical significance: How standard error costs us jobs justice and lives.* Ann Arbor: University of Michigan Press.

Chapter 4
About the Compatibility of the Diverging Interpretations

After having analyzed the different nature of $P(A_n)$ and $P(A_1)$ we are ready to discuss the relationships extant between these parameters.

4.1 Not Contradictory Approaches

The theorems of large numbers and of a single number explain the multi–fold behaviors of probability in the physical world, and demonstrate that these phenomena are not absolute truths but derive from the following hypotheses

$$n \gg 1, \tag{2.3}$$

$$n = 1. \tag{2.4}$$

The number n covers disjointed intervals and this separation yields that hypotheses (2.3) and (2.4) are mutually exclusive

$$(n \gg 1)\, OR\, (n = 1). \tag{4.1}$$

Accordingly, the probability of long term event and the probability of a single event can be used without conflict in point of logic

$$P(A_n)\, and\, P(A_1)\, are\, not\, contradictory. \tag{4.2}$$

The frequentist and subjective versions do not come into opposition both in the practice and in the theoretical plane because they rely on different situations. As relativistic theory does not deny classical physics since they are based on mutually exclusive hypotheses, so the frequentist and subjectivist notions cannot collide since they revolve on two distinct forms of event. One cannot calculate the two probabilities simultaneously as they regard distant circumstances. Therefore, working statisticians are free to select the most appropriate approach irrespective of any philosophical quarreling that declares reciprocal incompatibility.

P. Rocchi, *Janus-Faced Probability*, DOI: 10.1007/978-3-319-04861-1_4,
© Springer International Publishing Switzerland 2014

Howson and Urbach (1993) hold that physical and epistemic probabilities cannot be fundamentally distinct, and argue on the intellectual links which should be established between the two concepts.

The analytical approach which we are following, derives the significance of probability from $n \gg 1$ and $n = 1$. The present work assumes two distinct hypotheses and cannot share the philosophical perspective which relates the frequentist to the subjective interpretation and violates statements (4.1) and (4.2).

4.2 Method of Working

The abstract findings noticed above have practical implications. *The values $P(A_n)$ and $P(A_1)$ are logically distinct and both of them can be used in the professional practice.* This remark constitutes a cornerstone in the present theoretical framework and provides the guidelines for the use of the probability calculus in the working environment.

In a preliminary step, we examine more precisely the subjective method.

4.2.1 Lemma of Obligatory Subjectivism

Suppose z is any positive integer and the probability of A verifies this condition

$$P(A) = 1/z, \quad z > 0. \tag{4.3}$$

Then the relative frequency of the successful event A in n trials is not equal to the probability

$$F(A_n) \neq P(A). \tag{4.4}$$

If

$$1 < n < z. \tag{4.5}$$

Proof We proceed by absurd and deny (4.4) and impose the relative frequency equal to the probability

$$F(A_n) = N(A_n)/n = P(A) = 1/z. \tag{4.6}$$

If the event A occurs one time

$$N(A_n) = 1.$$

We obtain from (4.6)

$$1/n = 1/z.$$

That is

$$n = z.$$

This conclusion mismatches with assumption (4.5), thus (4.6) is false and (4.4) true.

As an example, let us figure out the probability of getting the ace of spades from a card deck

$$P(A_s) = 1/z = 1/52.$$

If the number of trials is less than 52, a priori we assert that the relative frequency cannot equal to $P(A_s)$. In other words, if $1 < n < 52$, then $P(A_s)$ cannot be controlled.

It may be said that the theorem of a single number establishes the lower bound of subjectivism and Lemma 4.2.1 specifies the upper bound for the impossible control of the probability. A practitioner has an obligatory way to follow when the number of trials ranges between z and 1; and one can conclude in the following way:

(I)   *The probability has necessarily a subjective significance as $z > n \geq 1$.*
(II) *The probability is a physical quantity as $n \to \infty$.*

Let us examine some practical consequences deriving from this couple of statements.

4.2.2 Mutually Exclusive Tactics

Sometimes working statisticians obtain very similar results when they use the Fisher/Pearson methods and when they adopt the Bayesian methods. Ambiguities emerging in statistical projects often raise the following questions:

As the two methods provide identical results: what is the best one?

Can we merge the methods in some way?

One can observe a variety of behaviors on the part of non–statisticians. The Bayesian standard appears to be simple and induces the generic owner of a project to give up classical, weighty statistics and to work with the new methods which seem to focus better on the specific target. Sometimes the purposes of the intended project do not appear very clear and the leaders tend to select the statistical modes which they themselves like. At last, we can see managers who choose the statistical approach which is less expensive.

Beside non–statisticians, we find a variety of behavior on the part of specialists.

Orthodox practitioners tend to declare the theory which they trust in, and later on use the related methods in the planned statistical work. Sometimes the member of a probability school seems like one who has embraced a religious faith, and forces the listeners to accept the results 'en bloc'. This line of conduct can be considered an arbitrary act in many respects. The personal adhesion to a theoretical framework does not match with the scientific method which operates beyond personal conviction. Individual credence should not influence a mathematical calculation.

Things do not go better in the pluralist camp. Theorists have not set up a rigorous theory so far, thus working statisticians miss precise criteria of discernment. They do not follow clear guidelines and frequently improvise in selecting the most appropriate method to follow. Experts make decisions on empirical basis and sometimes raise suspicions of individual judgment or convenience.

In summing up, personal preferences and arbitrary decisions are not so rare in companies and businesses, and may be found in some research sectors such as pharmacology (Chow 2011), meteorology (Robert 2008) and sociology.

In contrast to the multiform conduct of experts and non–experts in the working environment, the present theory yields a rather rigid reply to the previous queries.

Frequentism and subjectivism are based on mutually exclusive preliminaries

$$(n \gg 1)\, OR\, (n = 1). \tag{4.1}$$

In consequence of (4.1), one cannot tolerate any compromise or combination of statistical methods even when they provide results that are equal in numerical terms.

Some authors underline the technical differences between the two existing statistics (Kardaun 2005; Gill 2002). In the present frame, it is not a question of techniques; it is a question of substance. Hypotheses $(n \gg 1)$ and $(n = 1)$ are separated; $P(A_n)$ and $P(A_1)$ make great difference: the former is an authentic physical parameter, the latter is not extant in the real world.

Two precise ranges spell out the rigorous manner in which one should behave:

(I)   *The probability has necessarily subjective significance as $z > n \geq 1$.*
(II) *The probability is a physical quantity as $n \to \infty$.*

These rules do not leave any space to arbitrary or generic criteria. Practitioners should adopt the Pearson/Fisher methods when they investigate repeatable events (see point II); *practitioners should use the Bayesian statistics when they are concerned in single cases* (see point I).

In a way, the present theory matches with Gillies (2000) who claims there are two broad areas of intellectual study which require different interpretations of probability. The social/human sciences comply with the subjectivism and the natural/exact sciences need the frequentist techniques. Different interpretations fit better in different contexts. This is somewhat true and sharable, but guidelines I and II—conversely to Gillies—adopt the number of trials as the basic measure necessary to discern the school to follow and in principle, the number of experiments is independent of the environment. Long term events are not exclusive to exact science and it may happen that one has to adopt the Bayesian approach in quantum physics, and the frequentist one in sociology.

When the context appears fuzzy, the present theory suggests that one should clarify the goal to reach in advance of the statistical processes of calculus. For example, it is not sufficient to verify whether a new cure is effective and the leaders of the medical project should establish if the work is oriented toward a broad description or otherwise toward a criterion in favor of a single person. The alternative targets could be summarized in the following terms:

What is the probability of healing over x weeks using this new drug?

What is the probability of healing a patient who uses this new drug in x weeks?

The first query regards the general validity of the cure; the second query deals with a personal situation or single convenience.

Similar ambiguities emerge in various fields. For example, an expert in meteorology can be called to forecast the weather of a calendar–day in a specific zone or to optimize the algorithm which calculates the weather evolution in that zone. It is evident how the first project refers to A_1 and the second to A_n. One cannot travel along a third way and cannot refer to any middle path between the two statistical assumptions.

The separated ways are not so trivial to follow as they appear at first glance. Incompatible behaviors are required when one addresses a collective or otherwise a single experiment; $P(A_n)$ and $P(A_1)$ imply opposing strategies on the part of an agent. Take for instance the vexed conundrum:

Is it profitable to bet a number at lotto that has not hit for a long time?

The law of large numbers holds that frequency approaches probability as n increases. The longer the number has not been drawn, the more likely it is to be drawn in the next lotto game: sooner or later the *cold* number pays. Frequentists say 'yes' to the previous question. From another standpoint, the probability of a single event never changes. "Probability has no memory" for subjectivists and all numbers have the same chance of being drawn. Bayesians say 'no' to the previous question.

The present framework entails that both the answers are right but there is something more and above. Both the theories are correct but:

(1) The theories are alternative and there is nothing in–between them. This rule simply derives from the expression: $(n \gg 1)$ OR $(n = 1)$. In practice, a gambler has to select only one behavior. He has to state in advance whether to consider the general event or otherwise to consider single events, and cannot switch from one procedure to the other.

(2) The procedures have incomparable requisites. The subjectivist strategy has the *logic of finite* and in any moment the gambler can go on with the bet or abandon the lotto game. By contrast, the frequentist approach has the *logic of infinity*. The cold number will win 'for sure', but in order to recover the progressive losses the gambler has to undergo the following severe rules:

  – The gambler must never leave the lotto game in order to win or turn to the subjectivist strategy.

– The gambler must bet every time an amount of money sufficient to recover the overall losses thus he has to manage unlimited capital. Often the losses have an exponential trend.
– The manager of the game should permit this strategy and should have an infinite amount of money at his disposal.

In conclusion, the subjectivist approach is simpler and does not provide any special trick to win. The frequentist method seems to offer an ingenious expedient but has the logic of infinity which challenges the human capacities. Gamblers' thirst for money leads them to forget the above listed restrictions and the so called *gambler's fallacy* is the tendency to dream a big win and to forget the very large number of losses, and how those losses may add up to be significantly higher in value than the final win (Barron and Leider 2010).

References

Barron, G., & Leider, S. (2010). The role of experience in the gambler's fallacy. *Journal of Behavioral Decision Making, 23*, 117–129.

Chow, S. C. (2011). *Controversial statistical issues in clinical trials.* Boca Raton: Chapman & Hall/CRC.

Gill, J. (2002). *Bayesian methods: A social and behavioral sciences approach.* Boca Raton: Chapman and Hall/CRC.

Gillies, D. (2000). Varieties of propensity. *British Journal for the Philosophy of Science, 51*, 807–835.

Howson, C., & Urbach, P. (1993). *Scientific reasoning* (2nd ed.). Chicago: Open Court Press.

Kardaun, O. (2005). *Classical methods of statistics.* Heidelberg: Springer.

Robert, C. P. (2008). *The bayesian choice: From decision–theoretic foundations to computational implementation.* New York: Springer.

Chapter 5
Criticism on the Philosophical Pollution

Since the beginning of my inquiries, Popper's lesson fascinated me and I wondered why the propensity theory still raises debates and issues. I asked myself why Popper—in substance—missed his target in the probability theory whereas this giant of human thought achieved extraordinary results in other fields of research.

Why Popper and even other great authors did not find the definitive solution to the probability foundations?

I reached the conclusion that the method adopted by them constitutes the weak point of their theoretical constructions. I convinced myself that philosophy is not the appropriate mode to tackle the significance of probability. An issue pertaining to mathematics cannot be solved using a non–mathematical approach. I was persuaded that the various logical frames are conflicting and do not provide definitive answers because of a more or less intense philosophical pollution.

All of us agree that philosophers did a great deal of preparatory work. They tilled the ground of science and mathematics in such a way that they discovered intriguing truths, but—on my opinion—the definitive outcomes in the probability theory are waiting to be achieved by means of analytical methods which typically pertain to mathematics.

In the present chapter, I mean to explain the features of the present work to the reader and the contrast emerging with the current probability philosophies.

5.1 Analytical Assumptions

The inception of the present book declares that *analyticity* should be the matter of content rather than form. Any philosophical premise which risks becoming all–inclusive and providing fuzzy results has been placed apart.

In the first pages, two elementary hypotheses have been expressed in order to confront the problem of probability interpretation. One is able to demonstrate the

P. Rocchi, *Janus-Faced Probability*, DOI: 10.1007/978-3-319-04861-1_5,
© Springer International Publishing Switzerland 2014

two theorems and to pinpoint that probability behaves differently in different circumstances, thanks to the analytical assumptions

$$n \gg 1, \quad n = 1.$$

5.2 Separation of Areas

Dictionaries define the adjective *abstract* as "something that has been stated without connection to a specific instance"; "something considered apart from concrete existence". Abstract thinkers focus on the general properties of mathematical entities which they investigate, and the material meaning of those entities lies beyond the horizon of pure mathematicians. These ones do not have (or should not have) concern for the practical qualities of a mathematical expression. By definition, pure mathematicians should not pursue any specific purpose besides developing logic and reasoning skills.

Whereas a pure mathematician gives an abstract result without reference to a specific instance, a practitioner develops a calculation that describes a specific situation. Engineers, doctors, experimentalists and others use mathematical techniques to create concrete solutions in each field of interest. They pursue the purpose of gaining knowledge directed toward the production of useful devices, systems and materials. Practitioners particularize the meaning of the variables in each context and the concrete nature of the results emerges in their works.

From the first step, the present logical frame keeps the applied study of probability apart from the abstract study. This book assumes that Kolmogorov's axioms are true, and in explicit terms, recognizes the autonomy of abstract mathematicians who determine the general statements for $P(A)$. This work separates the speculative inquiries of pure mathematicians from the problem of testing probability by means of experiments.

5.3 Inclusiveness

A valid abstract mathematical theory should cover different domains of application, thus the present work does not have prejudices against any interpretation of probability and examines any possible scenario. This logical framework does not reject any hypotheses and follows two independent pathways. On one side, this book assumes that the number of trials is very large and demonstrates that probability is an authentic measure as necessary consequence. On the other side, this book considers only one trial and deduces that the probability can sustain a personal credence. Each inferential process starts with a precise hypothesis and arrives at the conclusions which conform to what first, von Mises and then de Finetti asserted.

In summing up, the methodology in question revolves around three special features:

(A) It follows the analytical mode of reasoning since the first stage.
(B) It keeps separated abstract issues on probability from applied inquires.
(C) It does not exclude any hypothesis which emerges from the living environment.

5.4 Terms of Comparison

In my opinion, the characters of the present approach should appear more evident if one examines points *A*, *B* and *C* in relation to the present–day theories. For the sake of simplicity, I confine myself to mention von Mises and de Finetti as the representatives of the two schools of thought.

(A) Commentators share the idea that the frequentist and subjective theories of probability start from the problem on how to interpret statements of probability. It may be said that von Mises begins with this conviction:

$$\textit{Probability is a real quantity.} \qquad (5.1)$$

De Finetti says:

$$\textit{Probability is subjective.} \qquad (5.2)$$

Each author has a special conception in mind and establishes a series of statements on the basis of (5.1) and (5.2) respectively. Von Mises explains how to obtain probabilities that can be controlled by experiments; de Finetti deals with probabilities which express personal credence. From the physical interpretation of *P*, von Mises lays down the notion of *collective* and utilizes the law of large numbers (LLN) to support his view. De Finetti assigns the subjective probability to single events so that LLN plays an ancillary role.

Each master introduces special principles and infers a sequel of mathematical results, all of them resulting from (5.1) or from (5.2). In substance, expressions (5.1) and (5.2) prove to be the real axioms of the frequentist and the subjectivist formulation in the order. Von Mises and de Finetti do not put these statements as official postulates but the entire logical constructions which they erect, come from these two propositions and thus one can but see (5.1) and (5.2) as authentic axioms.

The logical constructions created by von Mises and de Finetti demonstrate to be consistent in themselves. It may be said each edifice is solid inside its walls, but appears less strong outside; one can but note some questionable points. For example, a mathematical theory is usually grounded on elementary statements which produce results more or less complicated. A theory can even convey philosophical contents, as it proceeds from straightforward axioms toward complicated subject matters. Instead, von Mises and de Finetti follow the opposite directions; they begin with philosophical assumptions (5.1) and (5.2), and proceed toward mathematical formalization. A well–known aphorism says "Mathematics is not an opinion", but two noticeable probability theories are grounded on opinions.

A mathematical postulate expresses an elementary and self–evident truth since precision facilitates the control of the outcomes. In contrast, a philosophical assumption is very complicated in itself. The notion of subjective probability involves topics regarding the human knowledge, the personal psychology, and the context culture (Chiodo et al. 2004; Wallsten and Budescu 1983). Also, the concept of physical probability is not purely so self–evident and has some relations with the positivist philosophy (see Appendix A). Hence, the explanations given by von Mises and de Finetti are very difficult to handle because of their philosophical origin. Each author adds special remarks, rules and annotations to make his point of view more precise. They spill considerable amount of ink to demonstrate how their approach is reasonable and plausible but miss the target in the sense that their logical construction remains what they are: two philosophical frameworks.

By contrast, expressions $n \gg 1$ and $n = 1$ are self–evident from the mathematical perspective. There is no need to spend several words to justify these hypotheses.

(B) By definition, philosophy is the systematic inquiry into the principles and presuppositions of a field of study. A philosopher tends to address all the queries posited in the domain of interest and looks after all the debated issues. Von Mises and de Finetti share the style of thinkers, and confront the abstract issues on probability together and practical questions as well. They follow a special way: each master intends to create a general purpose theoretical construction and uses a unique notion of probability to reach his target. Von Mises and de Finetti superpose their own views of probability to the abstract concept of probability and in this way establish their second postulate. It may be said that every author fuses experimental probability and the mental concept of probability, and places this unique tenet in the spotlight. Using the present formalism, it may be said that von Mises posits

$$P(A_n) \equiv P(A), \tag{5.3}$$

On the other hand the probabilities $P(A)$ and $P(A_1)$ overlap in de Finetti's thought

$$P(A_1) \equiv P(A). \tag{5.4}$$

The two authors tend to convert an incomplete view into an account which should be applicable to all situations. To do so, each author presents preferably some of the facts while leaving out or minimizing other phenomena that are significant to interpreting and understanding the nature of probability.

Von Mises, de Finetti and others presume that one single model of probability should be used to confront any genre of problems and this decision reduces the effectiveness of the entire probability calculus. Principles (5.3) and (5.4) oversimplify the calculus and result in great difficulties. It is not a mystery that frequentists cannot handle single events; subjectivists obtain probability values that have problematic relationships with the physical reality.

The two schools of thought assume that the abstract and the applied calculations of P should overlap; but this simplified approach to probability contrasts sharply with standard scientific organization which places apart the sector of abstract mathematics from applied sectors.

What is worse from my personal viewpoint, each author does not justify his second postulate. Von Mises and de Finetti—to the best of my knowledge—do not clarify why one should melt together the abstract and applied versions of probability. They do not explain why one should adopt only one model of probability instead of many or what is the advantage of a unique argument for P and so forth. Identities (5.3) and (5.4) are used as two dogmatic postulates because of lacking justification.

By contrast, the present logical frame demonstrates that the sense of P is not intrinsic to P, but relies on the argument of P; and there is the abstract $P(A)$ on one side and the applied $P(A_n)$, $P(A_1)$ on the other side.

(C) Postulates (5.3) and (5.4) make a forced interpretation of the probability calculus, in that the former yields the denial of the single experiment probability

$$P(A_1) \text{ is false,} \tag{5.5}$$

And the latter rejects the calculation of long term events

$$P(A_n) \text{ is false.} \tag{5.6}$$

One is obliged to use either the aleatory or chance–based probability or alternatively the information–based or credence probability. The foisted split counters the evidence in the hands of working statisticians. The success achieved by traditional statistics and Bayesian statistics emerges *in rebus* (in things). Ordinary people sensed the prismatic meaning of P for a long. At last, the frequentist and subjectivist dogmatic behavior is contrary to the conclusions achieved by the theorems of large numbers and a single number which demonstrate that probability has double conduct in the world.

By contrast, the present logical framework takes into account any application. It does not mean to simplify the study of probability which includes a variety of subfields (we shall return to this topic later).

Concluding, I believe that the subjective and frequentist theories on the probability foundations present similar flaws. In my opinion, they both should be classified as philosophical works rather than mathematical frames as they rely on the following non–mathematical assumptions

$$Probability\ is\ a\ real\ quantity. \tag{5.1}$$

$$Probability\ is\ subjective. \tag{5.2}$$

Both the theories make smaller the probability calculus under the authority of these dogmatic decisions

$$P(A_n) \equiv P(A), \tag{5.3}$$

$$P(A_1) \equiv P(A). \tag{5.4}$$

Both of the authors oversimplify the multifold nature of probability as they respectively hold

$$P(A_1)\ is\ false, \tag{5.5}$$

$$P(A_n)\ is\ false. \tag{5.6}$$

Theoretical pluralists cannot unify or blend the current probability theories in a satisfactory manner because of the set of questionable assumptions from (5.1) to (5.6) just discussed. My conclusion does not go far from Burdzy's thought who claims that the modern philosophy of probability appears as a complete failure in our eyes.

5.5 Moderated Debates

It is commonly held that probability authors are good at pointing out the weakness of the others, but do not succeed in providing convincing answers about their own weak points. The situation is well described hereunder:

> The numerous, different, opposed attempts to put forward particular points of view which, in the opinion of their supporters, would endow Probability Theory with a 'nobler' status, or a more 'scientific' character, or 'firmer' philosophical or logical foundations, have only served to generate confusion and obscurity, and to provide well–known polemics and dis-agreements, even between supporters of essentially the same framework. (de Finetti 1970)

I perfectly agree that particular points of view, which have been sold as general solutions, generate confusion and endless debates. Dogmas $P(A_n) \equiv P(A)$ and $P(A_1) \equiv P(A)$ result in a variety of negative consequences. The superimposition of theoretical and applied studies negates—at least, in principle—the rational sub-division of labor which is commented in Sect. 2.4.2. Moreover the overlap of the probability notions opposes obstacles to the same proponents of the theories.

5.5.1 Death of Dogmatism

Authors who want that a particular interpretation—say $P(A_n)$ or $P(A_1)$—has the status of a comprehensive model, create very hazardous situations for their own followers. In fact, a general theory needs strong efforts to be prepared but *just one case is able to falsify a pretended general theory.* Commentators delight in devising paradoxes that undermine the opposing thesis. A sole exception can destroy the supposed general formulation.

I mean to comment on the criticism raised against the various probability theories which exclusively takes origin from the presumed status of 'general theory' assigned to each proposal:

- The formula which calculates the ratio of favorable and total cases offers a very effective assistance to working statisticians but draws severe problems when Laplace places this formula at the center of a supposed comprehensive framework. Appendix A briefly reminds how thinkers accuse the Laplace theory of logical circularity and inconsistency. A fine equation which people use and appreciate every–day becomes a blatant anomaly.
- Various aspects of the frequentist theory has been criticized (see Appendix A). The flaws find substance as long as von Mises pretends that his definition of probability is recognized as universal. Instead, if one accepts the frequentist scheme as an investigation which casts light on a practical facet of probability and does not assume the frequentism as the unique, universal reference, then the opposition against the frequentist view has no reason to subsist or becomes marginal. In fact, the present 'defects' become no more than the detailed 'conditions' to apply the large numbers law. For instance:

  - The two principles of von Mises (see Appendix A) specify the features that a sequel of random independent events should have.
  - The definition of *collective* helps statisticians to determine the set to measure.
  - The notion of limit becomes a conceptual reference rather than the analytical operation of limit, and so forth.

  If a frequentist is not involved in useless titanic efforts to demonstrate that the law of large numbers should be used in all situations, he merely employs the theorem as long as the theorem assumptions are true.

- De Finetti underlines that the subjective philosophy can apply to all the phenomena related to probability whereas the same cannot be said of the frequentist one. Wald (1950) claims that every admissible statistical procedure is either a Bayesian procedure or a limit of Bayesian procedures. Subjectivists mean to demonstrate that their theory is *universal* on the basis of practical evidence.

  To settle this dispute, we should examine the term *'universal'* which denotes:
  1st—An important concept.
  2nd—Something broadly adopted.

Sometimes the two meanings overlap, other times they cannot be simultaneously true. For example, the third principle of thermodynamics states that the Boltzmann entropy S_B of a system is typically zero at absolute–zero temperature

$$S_B = k_B \ln N = 0.$$

This principle has universal import since it explains a fundamental property of gas statistics but is rarely used in applications. The first meaning of the adjective 'universal' is true, the second is false.

Subjective probability proves to be extremely flexible for the simple reason that all the random events can be tackled one by one. A lot of situations in economy, in physics and in various environments can be addressed individually in succession (Carlin 2008). Decision makers are more inclined to cope with a single situation rather than to manage a long term strategy. The 2nd meaning reported above is true.

The theorem of a single number demonstrates that the probability of a single event is not real, and subjectivists would sustain the superiority of the symbol $P(A_1)$ respect to $P(A_n)$ when the latter is a real quantity, and the former is a remedy to the lack of a realistic measure. The subjectivist school would convince people to give up any realistic forecast and in its place take a quantity which "does not exist". This gigantic effort comes up against the desires of scientists, professionals and managers who are usually concerned in realistic data. Nothing is more important for experts than the likelihood of a mathematical outcome, no matter what they pay for it. Therefore, the probability $P(A_1)$, which nobody can corroborate, cannot be judged predominant over the physical measure $P(A_n)$ that can be confirmed or disproved. The 2nd meaning of the adjective 'universal' is true for the subjective probability; but the 1st meaning is false since the subjective probability is not a scientific measure in full sense of the terms.

If Definettians and Bayesians relinquish their proud claims and put aside the non–demonstrated statement $P(A) \equiv P(A_1)$, then the controversy expires. The subjectivist view becomes an amazing idea to circumvent the difficulties raised by single event probability, and people find support from the Bayesian methodologies without obstacles.

As soon as one gives up the status of universal and comprehensive theories for the frequentist and subjective ones, most of the conceptual criticism raised against them vanishes.

5.5.2 Empirical Problem Solving

The present theoretical framework places a division between the abstract and the applied calculus of probability; hence one should tackle several questions on the basis of pragmatic criteria. Some vexed riddles treated as 'general issues' so far can be addressed using practical methods. I mean to exemplify this style of work with the following cases.

- Engineers are aware that the calculated value y_c and the corresponding measured value y_m are never identical; and the difference between y_c and y_m can be accepted as long as it falls within the range established by standard procedures. This range has different size in various contexts. For example, precision mechanics tolerates widths in the order of hundredths of millimeters, civil engineering in the order of millimeters, topography in the order of decimeters, and astronomy in the order of kilometers.

In the present book the distance between the empirical probability $F(A_n)$ and the calculated probability $P(A_n)$ can be seen as a limit comparable to the difference emerging between y_c and y_m in other sectors. Factually, working statisticians operate in a manner similar to engineers and professionals; they select the likelihood of the results in accordance with the requirements of the specific project.

- Practical needs suggest employing an integer in the place of a decimal number. For instance, an economic amount is commonly rounded up or rounded down in order to make easier the payments; the decimal number is lowered or raised to the nearest whole number. Real needs provide motives for rounding up or rounding down probabilistic values too. To exemplify, a doctor considers negligible the risk of the infection spreading from one person to another when the probability of contacts is close to zero. On the other hand, when the probability is very high the event is deemed as certain. For example, when the probability of tornado striking a certain area is high, local administrators assume the tornado will arrive for sure and arrange all the needed countermeasures.

Cournot (1843) established that a random event with very small probability is morally impossible; it will not happen. This statement ranked the status of 'principle' in literature. Paul Lévy (1925) considered it—together with the Bernoulli theorem—as the main bridge to the empirical world. Castelnuovo (1933), Hadamard (1922), Kolmogorov and others recognized the Cournot criterion as a 'fundamental law of chance'.

There is no widely accepted scientific basis for claiming that certain positive values are universal cut–off points for effective negligibility of events. But the present framework does not need to invoke any 'principle' and shares the pragmatic comment that Borel made in a couple of popular non–technical books written late in his life (Borel and Scott 1963; Borel 1962). He estimates 'practically impossible' and roundable to zero, the probabilities less than 10^{-6} on the human scale, the probabilities less than 10^{-15} on the earth scale, and less than 10^{-50} on the cosmic scale and, lastly, he considers irrelevant the values less than 10^{-1000} on the universal scale.

- The aphorism "Probability has no memory" fascinates several experts who claim anything may happen. There is no force, which at some point in an infinite sequence of events, will eliminate the probability of coming up black and force it to come up red. Thus, theorists accept that the roulette may generate an indefinite sequence of the same number or may output alternate red and black numbers. These remarks appear correct in abstract beyond any doubt but turn out to be somewhat absurd in the practice. "Probability has no memory" is a sort of theoretical myth which shows blatant differences when compared with the living environment.

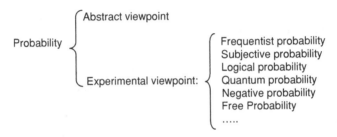

Fig. 5.1 Abstract and applied studies of probability

If one assumes that an event is random, he expects that the outcomes of that event create a disordered series in the real world. Instead, if the series of outcomes is regular, he concludes that the initial assumption is wrong and immediately does something about it. In organizations, repeated controls are carried out for this purpose and an episode can illustrate this managerial solicitude.

Early in the 20th century the engineer William Jaggers had thoroughly examined the roulette mechanics and noticed its rotation was guided by a cylinder with a cotter pin on top (Smith 2011). This was the weak spot of the roulette revolution since an imperceptible wear of this cotter pin unbalanced the system and the results were no longer equally likely. For over a month he registered the outcomes from the roulette tables in the Montecarlo Casino. He pointed out the device producing preferential numbers by analyzing his data and won two million and four hundred thousand francs in few days. This amount—extraordinary at those times—alarmed the Casino's management who discovered the causes and immediately eliminated them.

The idea that the roulette can come up the same number every time sounds like a joke.

5.6 Open Doors

The theorems of large numbers and of a singe number provide exhaustive answers upon the experimental control of the calculated values of probability. The number of trials n can vary from one to infinity, and assumptions $n \gg 1$ and $n = 1$ cover the entire axis n.

Does the present logical frame exclude other interpretations of probability?

The two theorems answer thoroughly the controllability problem but do not deplete the applications of probability (Fig. 5.1).

I return back to the case regarding the infinitesimal calculus. The calculus of derivatives in mechanics—introduced by Newton—is very significant but is not exclusive to the mechanical field, it serves several application areas. In a similar manner, there are many application fields of probability. The experimental control does not exhaust the study of *P*. For example, Carnap means to measure the confirmation of a formal hypothesis using probability. Quantum physicists forecast the behavior of atomic particles exploiting statistics in a systematic manner (Accardi 1981). *Negative probability* offers support to a number of physical and economic problems (Haug 2007); *free probability* is an area of research focusing on non–commutative random variables; *differential probability* is used in cryptology (Lai et al. 1991) and so forth.

Each group of probabilists treats a precise typology of applications and draws attention to a special kind of argument. I guess an even larger number of studies will enrich the family of applications as time passes.

References

Accardi, L. (1981). Topics in quantum probability. *Physics Reports, 77*, 169–182.

Borel, É. (1962). *Probability and life*. New York: Dover.

Borel, É., & Scott, D. (1963). *Probability and certainty*. NewYork: Walker.

Carlin, B. P., & Louis, T. A. (2008). Bayesian Methods for Data Analysis-Chapman and Hall/ CRC.

Castelnuovo, G. (1933). *Calcolo delle Probabilità*. Bologna: Zanichelli.

Chiodo, A. J., Guidolin, M., Owyang, M. T., & Shimoji, M. (2004). Subjective probabilities: Psychological theories and economic applications. *Federal Reserve Bank of St. Louis Review, 86*(1), 33–48.

Cournot, A. (1843). *Exposition de la Théorie des Chances et des Probabilités*. Paris: Hachette.

De Finetti, B. (1970). *Teoria della Probabilità*. Torino: Einaudi. Translated as *Theory of probability*. New York: John Wiley and Sons (1974).

Hadamard, J. (1922). Les Principes du Calcul des Probabilités. *Revue de Métaphysique et de Morale, 29*(3), 289–293.

Haug, E. G. (2007). *Derivatives models on models*. New York: John Wiley & Sons.

Lai, X., Massey, J. L., & Murphy, S. (1991). Markov ciphers and differential cryptanalysis. *Proceeding of 10th International Conference on Theory and Application of Cryptographic Techniques*, pp. 17–38.

Lévy, P. (1925). *Calcul des Probabilités*. Paris: Gauthier–Villars.

Smith, G. (2011). *Essential statistics, regression, and econometrics*. San Diego: Academic Press.

Wald, A. (1950). *Statistical decision functions*. New York: Wiley.

Wallsten, T. S., & Budescu, D. V. (1983). Encoding subjective probabilities: A psychological and psychometric review. *Management Science, 29*(2), 151–173.

Part II
Considerations Around the Probability Axiomatization

Chapter 6
Some Remarks on the Argument of Probability

The first part of this book focuses on the available methods for testing probability and in turn the frequentist and subjective interpretations were commented on. The previous pages try to show how the difficulties of a working statistician have a tendency to vanish when he rejects the philosophical theories and confront the issues through the analytical approach.

There are still some problems remaining for probabilists.

6.1 Abstract Axioms

Kolmogorov held that his system of axioms is *consistent* but added that it is *incomplete* because of the variety of problems that emerged—and indeed still emerge—in consequence of new statistical applications. The construction of the Soviet mathematician is correct due to its inner consistence, but there are territories that do not seem to be covered by the axioms and some thinkers are questioning the Kolmogorovian axiomatization.

De Finetti believes that the series of phenomena and rules considered in the subjective theory go beyond a purely mathematical framework. Even so, a group of subjectivist writers attempt to make more precise the connections between quantitative probability and the coherent personal dispositions toward uncertainty (Fishburn 1986). The von Neumann–Morgenstein axioms refer to comparison among utilities when the agent measures his beliefs with finitely additive probabilities. Savage (1954) accepts the so–called *axioms of rationality* that are the principles of *transitivity, dominance, non–influence of the formalization,* and *symmetry*; he also appends the fifth principle that is *independence* or *sure thing axiom* to ensure rational decision under uncertainty. Cox (1946) formulates the quantitative rules of plausible inference in relation to the Bayesian probability. Logicians—such as Carnap, Wright, Roeper and Maher—present appropriate assumptions for inductive probability; however, their proposals differ in some respects.

P. Rocchi, *Janus-Faced Probability*, DOI: 10.1007/978-3-319-04861-1_6,
© Springer International Publishing Switzerland 2014

Von Mises bases his definition of probability on two principles. The former fixes the global regularity of the collective; the latter assumes that the collective has local irregularity.

Popper (1959) provides a complex elaboration of the probability postulates. In particular, the preliminaries of his theory comprehend *Group A* which is a set of statements that are practically an adaptation of the postulates for the so–called *algebra of logic* (Huntington 1904), and *Group B* which gives the axioms peculiar to the measurement of probability.

Among the most recent contributions, Burdzy (2009) puts forward five 'laws' for making probability assignments. He points out that axioms merely encode facts which we have to regard as uncontroversial. Laws instead, are proposals for a scientific theory that are open to falsification.

In the close, I mention the various attempts to define negative probabilities (Curtright and Zachosy 2001) and non–standard probabilities (Borba de Andrade 2007).

I am not sure whether we should keep the Kolmogorov axioms, or should modify them, or even if we should follow another course to establish the postulates of the probability theory. I do not see these kinds of issues so easy to treat; and imagine an alternative maneuver in place of an attack on the problem of axioms in a direct manner.

Axioms are the foundations of a formal deductive system, and are normally used to make very specific predictions. If the predictions are generic or wrong, then the axioms would be improved or rejected. But to ameliorate a mathematical axiom, one often revises some basic notions; and injects more precise concepts.

One could recommend the specific description of the intended objects to calculate in order to make better the probability axioms and in particular *one could pay greater attention to the random event modeling*. I am inclined to conclude that the probability argument—that is the assortment of variables to which P refers—should be the staple of probability foundations.

I am aware that my opinion on the importance of the probability argument is not so popular; hence I mean to justify my personal position.

6.2 System Modeling

In the first part of this book the concepts A_n and A_1 played leading roles. We have explored the use of P in the world on the basis of the random event's form. The second part of the book searches for accurate modeling of random events:

Why this insistence on the argument of probability?

What are the origins of this pressing attention?

My initial field of interest was software engineering and I particularly investigated the software analysis techniques. This topic is rather specialist and perhaps a few details will be useful to the reader who is not fully cognizant of software analysis and design (Fig. 6.1).

Fig. 6.1 Context DFD (*top*), and Application DFD (*bottom*) of a software program used by vendors, customers and users

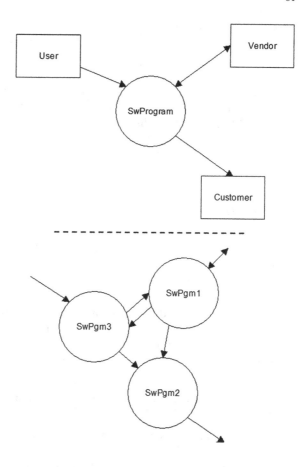

6.2.1 *DFDs*

In the early seventies, software developers raised the issue of system representation. They needed suitable methods for describing the functional components of software applications and arose over this problem by introducing the *multi–level graph* that provides multiple images of a single system.

The *Data Flow Diagram* (DFD) proposed by Larry Constantine and Edward Yourdon became a standard in software engineering in the eighties (De Marco 1978). DFDs normally exhibit a system in three levels of zoom. The first scheme, called *Context DFD,* depicts the overall arrangement; it exhibits the software package in the centre and all around the entities interacting with it. The *Application DFD* explodes the bubble placed in the centre of the Context diagram. The *Work DFD* provides further insight into the bubbles of the Application diagram. The three graphs render different levels of visual granularity to the system. Special symbols help the viewer to relate an image to another one.

Fig. 6.2 Entity–relationship
diagram

Multilevel analysis of systems progressed from theoretical and applied view-points (Parisi and Piersanti 1995). Researchers optimized the methods in order to cluster and to partition graphs (Boutin et al. 2005). These techniques have demonstrated a high flexibility and have spread into a variety of fields such as ecology, electronic chip design and operational research.

6.2.2 ERD

In the seventies, Peter Chen (1976) devised the *Entity–Relationship Diagram* (ERD) which came into general use in relational database design. A database can archive several types of data, and the ERD can describe a variety of situations regarding the archived data. In a nutshell, the boxes include the subject or the object of a sentence and the diamond includes a passive or active verb. Each branch of the graph exhibits a sentence that makes comprehensible on how data conform to the physical reality (Fig. 6.2).

Peter Chen forecasted the broad applicability of his graph:

> The entity–relationship model adopts the more natural view that the real world consists of entities and relationships. It incorporates some of the important semantic information about the real world. (Chen 1976)

Chen's model matches with the thoughts of classical philosophers, such as Plato and Aristotle, who associated human knowledge with the apprehension of forms and their relationships to one another. Modern theorists show how the elements ER offer the essential alphabet to codify the world in a natural manner (Kent 1984).

6.3 A Negligible Topic?

I gave my contribution in optimizing the use of DFDs and ERD, and became convinced of the virtues of these techniques which are capable of representing complicated situations very accurately. When I migrated from software engineering to the probabilistic field, it was natural for me to observe how probability theorists pay little attention to the description of indeterminate phenomena. Whereas software designers spent much energy in analyzing systems, I found out that mathematicians had scarce concern for event modeling. This inclination was not a recent one and had deep roots. Researchers deemed the probability argument a negligible topic for centuries.

A brief historical excursus can elucidate my opinion (Rocchi 2006).

6.3.1 Famous Correspondence

The involvement of Pascal with the probability calculus may be placed inside a very short interval of time, approximately between the years 1651 and 1654. During this period the French mathematician became close to his childhood friend, the duke of Roannez, who in turn presented Antoine Gombaud, better known as chevalier de Méré.

Gombaud posed an interesting question to Pascal; in particular he believed that the probability of getting at least one six with 4 throws of a single die was the same as the probability of getting at least one double six in 24 throws of a pair of dice. Pascal proved his friend's reasoning is false as the bet is favorable after 25 throws when the probability of throwing a double six is greater than 1/2 (Ore 1960).

Shortly after de Méré placed a second query about the proper division of the stakes when a game of chance is interrupted. Pascal methodically began by classifying all the situations under the constraint that only one game is missing for the final win. Then he examined when two games were missing; and so forth. Blaise Pascal achieved the solution to the so-called 'problem of points' but this was unsatisfactory for him.

Pascal was in correspondence with Pierre Fermat, who was from Toulouse, and decided to contact him in order to discuss the 'problem of points'. Fermat briefly mentioned a solution but did not make the reader aware of the approach he followed. Pascal expressed his doubts about this partial answer, but refrained from a complete manifestation of his thoughts; perhaps for not hurting Fermat's feelings. Later Pascal went back to the 'problem of points' and extensively demonstrated his queries. A brief reply arrived a few days later from Toulouse as though the writer meant to assure his friendship rather than to deal with the query. Nearly a month later, on the 25th of September 1654, Fermat definitively justified his calculations and Pascal, perfectly satisfied, thanked his colleague (Table 6.1).

The two mathematicians approached the problem through two distinct ways: Fermat suggested the use of permutations whereas Pascal employed a method roughly resembling the calculus of finite differences (Boyer 1968). They reached the same correct conclusions through different roads and these independent achievements matured Pascal's thought. The calculus of chances was no longer a resource for gamblers, but emerged as a distinguished mathematical sector from Pascal's viewpoint. We have the evidence of his belief thanks to a fortuitous event.

6.3.2 Pascal's Conjecture

In the first half of the seventeenth century a number of intellectual clubs were set up in Paris on the model of Italian circles such as the *Accademia dei Lincei* founded in 1603, which is still active today. Various savants such as Henri–Louis

Table 6.1 Synopsis of the letters exchanged by Blaise Pascal and Pierre Fermat		Date
	Fermat → Pascal	Undated, 1654
	Fermat → Pascal	Undated, 1654
	Fermat ← Pascal	Wednesday, July 29, 1654
	Fermat ← Pascal	Monday, August 24, 1654
	Fermat → Pascal	Saturday, August 29, 1654
	Fermat → Pascal	Friday, September 25, 1654
	Fermat ← Pascal	Tuesday, October 27, 1654

Habert de Montmor, Melchisédech Thévenot and the abbots Bourdelot and Mersenne, were bunched into small, informal groups of scientists and thinkers (Brown 1967).

In particular, Mersenne became the founder of *Academia Parisiensis*, which brought together scholars from various backgrounds. He made a careful selection of the scholarly persons who met at his convent in Paris or who corresponded with him from all over Europe. François Le Pailleur, son of a royal official became director of this academy after Mersenne and in 1654 asked Pascal for a report of his scientific quests.

Pascal's report, besides a very formal introduction, recounts an impressive list of topics such as numerical analysis, magic numbers, geometry of conics, geometry of plain loci and solids, theory of perspectives and searches on the vacuum. The letter closes with the following extract:

> Ambiguae enim sortis eventus fortuitae contingentiae potius quam naturali necessitati meritó tribuuntur. Ideó res hactenus erravit incerta; nunc autem quae experimento rebellis fuit rationis dominium effugere non potuit. Eam quippé tantá securitate in artem per Geometriam reduximus, ut certitudinis eius particeps facta, jam audacter proteat; & sic matheseos demostrationes cum aleae incertitudine jungendo, et quae contraria videntur conciliando, ab utraque nominationem suam accipiens, stupendum hunc titulum jure sibi arrogat: Aleae Geometria. (Chevalier 1950, pp. 70–71) (In fact the events caused by ambiguous chance are rightly credited to the contingent fortune rather than to the physical necessity. As a consequence the knowledge of random events roamed hesitant so far, nowadays instead those facts rebel to the experiment cannot escape to the rule of the rationality. And, thanks to the geometry, we have treated those random phenomena by exact calculus which shares the mathematical certainty and is just progressing in a daring way; and by joining the rigor of scientific demonstrations with the uncertainty of chances and by reconciling these two apparently contradictory items, we can assign the following amazing title: The Geometry of Chance to the discipline embracing both of them)

Rigorous tractability of the chances through the mathematical language, the elegance of distinct methods, the precision of calculus and the conformity to the arithmetic triangle convinced Pascal that probabilistic issues pertain to the scientific realm, and in addition, give body to an autonomous mathematical sector. The scientist wrote a few lines or only a few words for each subject of research piloted by him in those years; by contrast he devoted nearly a third of his letter to probability. This broad dimension reveals the fervor of the French mathematician for the discovery of this new fascinating sector. Pascal is recognized as the father of the modern probability calculus not just because he solved two problems but because of his awareness of the new territory of study which was opening.

The report to the Pailleur academy brings out the author's philosophical vision and in addition unveils a precise mathematical conjecture. The section devoted to the probability calculus closes with this significant statement:

> We can assign the following amazing title: The Geometry of Chance to the discipline embracing both of them.

Dominique Descotes (2001) has recently analyzed the French word '*geometrie* (geometry)' which has two very different meanings in Pascal's production. On one side, he uses 'geometrie' as synonymous with 'mathematics', 'rationality' and 'logic' in specialized works. On the other side, 'geometrie' has broad significance in philosophical contexts. Take for example his famous meditation on '*l'esprit de geometrie*', which regards the rational aptitude of man, in contrast with '*l'esprit de finesse*', which indicates the emotional side of the soul. By contrast, the word 'geometrie' has its real significance in technical reports, where Pascal uses 'geometry' to denote 'the study of the space and of its elements' in scientific papers (Magnard 2001).

The letter to the Academy recounts Pascal's researches and the use of Latin confirms it as a rigorous, scientific report. It is evident that the Latin word '*geometria*' means 'geometry' in the strictest sense of the word and in turn '*Aleae Geometria*' means 'Geometry of Chances' verbatim. Pascal considers 'Aleae Geometria' as the strict application of geometry to random events; 'Aleae Geometria' is the study of spaces and points in the probabilistic area. The author of the letter has in mind that the new mathematical sector can qualify the chances with precision and so *the chances are geometrical points*. The French author hints a precise conjecture on the probability argument even if he does not develop his hypothesis formally. *The random events should be assumed as points from the mathematical viewpoint.* The riddles solved by Pascal consolidate this interpretation. In fact he identified the outcomes of gambles by combinations and possibilities, which he took as elementary entities. He derived the probability by counting methods and evidently the chances can only be points.

In those years and during most of the next decades, Christian Huyghens, Jakob Bernoulli, Abraham de Moivre and others undertook matters basically related to gambling (Feller 1971). They focused on bets, craps, stakes and gambles, and calculated the 'possibilities of winning money'. The Pascal conjecture tacitly sustained the probability calculus but the argument of the probability was not yet worthy of definition on the theoretical plane.

First, Laplace (1814) tackled the methodical study of the probability notion. His 'Théorie Analytique des Probabilités' put forward a precise description of probability in the following terms:

> La théorie des hasards consiste à réduire tous les événement du même genre, à un certain nombre de cas également possibles, c'est-à-dire, tels que nous soyons également indécis sur leur existence; et à déterminer le nombre des cas favorables à l'événement dont on

cherche la probabilité. Le rapport de ce nombre à celui de tous les cas possibles, est la mesure de cette probabilité qui n'est ainsi qu'une fraction dont le numérateur est le nombre des cas favorables, et dont le dénominateur est le nombre de tous les cas possibles. (The theory of chance consists in reducing all the events of the same kind to a certain number of cases equally possible, that is to say, to such as we may be equally undecided about in regard to their existence, and in determining the number of cases favourable to the event whose probability is sought. The ratio of this number to that of all the cases possible is the measure of this probability, which is thus simply a fraction whose numerator is the number of favourable cases and whose denominator is the number of all the cases possible)

$$P = \frac{n_F}{n_P}.$$

The count of n_F and of n_P entails the idea that random events are nothing other than points but the Laplacian framework still left the Pascal conjecture unexpressed. Laplace himself called 'analytical' his conspicuous work nonetheless he refrained from fixing the argument of probability in analytical terms.

In the nineteenth and the twentieth centuries applications expanded in such widely different fields as physics, chemistry, economics, engineering, sociology and psychology. Different schools contributed to the advancement of the probability calculus that became even richer in contents but the Pascalian concept of the probability argument still remained as a secondary topic.

A turning point occurred when Kolmogorov (1950) built up the probability theory on the basis of precise axioms and wrote:

> Let E be a collection of elements ξ, η, ζ..., which we shall call *elementary events*, and $\mathfrak{S}$ a set of subsets of E; the elements of the set $\mathfrak{S}$ will be called *random events*.

Finally, the random event was worthy of attention and was defined as a subset in the event space. Kolmogorov sanctioned the Pascal conjecture *after nearly three centuries*, a period which seems enormous to modern sensibility but, in my opinion, illustrates the scarce consideration paid by the probability theorists for this topic.

References

Borba de Andrade, B. (2007). Topics in nonstandard probability theory. *UMI Dissertations Publishing*.

Boutin, F., Thièvre, J., & Hascoët, M. (2005). Multilevel compound tree. *Construction, visualization and interaction, Lecture Notes in Computer Science* (pp. 847–860). Heidelberg: Springer.

Boyer, C. B. (1968). *A history of mathematics*. New York: Wiley & Sons.

Brown, H. (1967). *Scientific organizations in seventeenth century france (1620–1680)*. New York: Russell & Russell.

Burdzy, K. (2009). *The search for certainty: on the clash of science and philosophy of probability.* Hackensack: World Scientific.

Chen, P. (1976). The entity-relationship model: Toward a unified view of data. *ACM Transactions on Database Systems, 1*(1), 9–36.

Chevalier, J. (Ed.). (1950). *Oeuvres Complétes de Blaise Pascal.* Paris: Gallimard.

Cox, R. T. (1946). Probability, frequency, and reasonable expectation. *American Journal of Physics, 14,* 1–13.

Curtright, T., & Zachosy, C. (2001). Negative probability and uncertainty relations. *Modern Physics Letters A, 16*(37), 2381–2385.

De Marco, T. (1978). *Structured analysis and system specification.* New York: Yourdon Press.

Descotes, D. (2001). *Blaise Pascal, Littérature et Géométrie.* Presses Universitaires Blaise Pascal, Clermont–Ferrand.

Feller, W. (1971). *An introduction to probability theory and its applications.* New York: Wiley.

Fishburn, P. C. (1986). The axioms of subjective probability. *Statistical Science, 1*(3), 335–358.

Huntington, E. V. (1904). Sets of independent postulates for the algebra of logic. *Transactions of the American Mathematical Society, 5,* 288–309.

Kent, W. (1984). Fact–based data analysis and design. *Journal of Systems and Software,* special issue, *4*(2–3), 99–121.

Kolmogorov, A. N. (1950). *Foundations of probability theory.* New York: Chelsea.

Laplace, P. S. (1814). *Théorie Analytique des Probabilités* (2nd ed.). Paris: Courcier.

Magnard, P. (2001). *Le Vocabulaire de Pascal.* Paris: Éditions Ellipses.

Ore, O. (1960). Pascal and the invention of probability theory. *American Mathematical Monthly, 47,* 409–419.

Parisi, F. & Piersanti, G. (1995). Multilevel graph grammars. In *Proceedings of the 20th International Workshop Graph Theoretic Concepts in Computer Science* (pp. 51–64).

Popper, K. R. (1959). *The logic of scientific discovery.* London: Hutchinson & Co.

Rocchi, P. (2006). *De Pascal à Nos Jours: Quelques Notes sur l'Argument A de la Probabilité P(A).* Actes du Congrès Annuel de la Société Canadienne d'Histoire et de Philosophie des Mathématiques (CSHPM/SCHPM) Toronto, *19,* 228–235

Savage, L. J. (1954).*The foundations of statistics.* New York: Wiley & Sons.

Chapter 7
Classical Modeling of the Probability Argument

The inattention of mathematicians to the argument of, probability goes on in the present day. Theorists are inclined to elude the random event analysis so far. For many, the probability theory becomes just a set-measurement theory and the argument of probability could be called an abstract and rather negligible detail in this context (Kallenberg 2002).

Is this opinion undisputable?

The praxis of science offers a lesson in this regard. Almost always scientists define what they want to process and in a second stage they measure the intended elements. Normally researchers begin with the objects and the phenomena which they aim to investigate and later devise the most suitable measurements for these arguments. It could be mentioned the electrical sector. By the end of the eighteenth century pioneer scientists found out electrical items and afterwards ascertained their qualities by means of appropriate measurement units.

The probability calculus makes an exception to this rule. Under the pressure of practical needs, theorists explored this field while the elements qualified by probability laid in the background. For three centuries those elements were neither formalized. One could wonder:

If we do not know well the object which we measure by means of probability, can we have clear ideas on the probability itself?

Authors usually assign a probability value to a random event: are we able to describe the random event in an appropriate manner?

7.1 Subset

For Kolmogorov the *random event is a subset* and the probability is a measure of sets. Let us make some remarks on this representation of the probability argument:

- Kolmogorov (1933) stated his axioms and soon after spelt out their practical application. He used with indifference the subset A for the *random event* and for

P. Rocchi, *Janus-Faced Probability*, DOI: 10.1007/978-3-319-04861-1_7,

Fig. 7.1

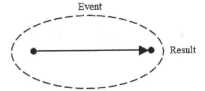

its *result* since the first pages of *Grundbegriffe*. Several scholars share this double designation which however causes some perplexities.

If one talks about 'event' and 'result', he tacitly admits that he is dealing with a complex physical situation which includes a number of components. In fact, the term 'event' denotes the entire occurrence and 'result' denotes the final outcome of a dynamic action (Fig. 7.1). It is evident that the properties of the event—which is the *whole*—are quite different from the properties of the result—which is a *part*—and theorists should keep them clearly distinctive one from the other. One could ask:

Q.1—How can we justify the set model for the entire event and/or for a part of it?

• The majority of authors who label the random event using its result, assume that the result will appear for sure, but this is misleading. A large number of events do not output any item, or the final state of the observed object cannot be detected, or one cannot get a precise result. Take this example: Mr. Smith suspects Mr. Thomson is unethical. Smith is silent; the suspicion lies inside his mind and does not generate anything. Since there is no result, the set model A cannot be applied to it. The reader should not deem silent events as trivial cases; cancer offers a noticeable example of a silent disease.

A second group of events emits outcomes although these prove to be insignificant for the purpose of identifying the event from which they have originated. For example, the crash of an airplane, and the disastrous fire of a plant require accurate analysis because the total destruction does not qualify the event. Two fires seem identical if we compare the final effect, while they are completely different if we analyze the causes and the internal dynamics of the fires (Vose 2000). All this motivates the following query:

Q.2—How can we calculate the probability value if the set does not appear in the world or cannot identify the random event?

• Sets seem pervasive because of the high abstraction of the set theory; instead, when we leave gambling games and some other special events, the set model creates difficulties. Several outcomes cannot be given as sets in statistical applications. Quantum physics brings forth an exception of significant importance (Gilson 2013).

Fig. 7.2 Single slit pattern

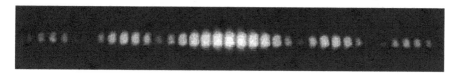

Fig. 7.3 Double slit pattern

In the two slits experiment a coherent light source illuminates a thin plate
pierced by two parallel slits. The light passing through a single slit strikes the
screen and makes a pattern that is easily interpreted as set A (or B) (Fig. 7.2).
When there are two open slits, the resulting pattern consists of bright and dark
bands on the screen. The two beams make an interference spectrum (Fig. 7.3).
One cannot calculate the conditional probability for the simple reason that the
final pattern presents variable density. The result cannot be modeled as intersection $(A \cap B)$ and the following equation is nonsensical

$$P(A \backslash B) = \frac{P(A \cap B)}{P(B)}.$$

The set model does not depict the final result (Feynman and Hibbs 1965).

Q.3—Why does the set model fail?

Several researchers are seeking the physical causes of the two slits experiment.
No matter whatever solution will be found in quantum mechanics, experimental
evidence shows that the set model does not work appropriately in all situations.

7.2 Sentence

Subjectivists and logicians who openly declare to address complex phenomena
adopt the linguistic representation of the random event, and hold other schemes to
be reductive. In general, many propositions can describe the same situation. Let the
propositions $p, q, \ldots$ describe an event; they are equal in meaning and belong to the
equivalence class A (Green 1988) that constitutes the model of the random event

$$A = \{p, q, \ldots\}.$$

• Broad literature brings evidence of the ambiguity of natural language. A text, a sentence or a single word can have a sense which is susceptible to different interpretations. Keynes and Carnap notice that even the term 'event' is misleading; this word has two common meanings as it can denote 'the events of a certain kind' (here symbolized by A) and also 'a specific event' (here A_1).

Languages give origin to countless forms of equivocal and inaccurate communication. Experts sum up this immense matter into three major kinds of equivocation:

– Vague or ambiguous statements.
– Statements that are not literally false but that cleverly avoid the truth.
– False statements by means of unreliable terms.

An agent can involuntary introduce an equivocal term in the description of the random event, or can create an incomplete description. The famous question posed by Laplace: "What is the probability that the sun will rise tomorrow?" offers an example of sketchy account. One wonders:

• Where will the sun rise? The sun will never rise in a deep coal mine.
• From which side of the horizon will the sun rise? It will never rise in the west.
• At which latitude will the sun rise? The sun does not rise at the poles for many months every year.

Natural languages do not furnish accurate accounts and some objections raised against Bayesianism, in reality express opposition to linguistic misleading; one concludes:

Q.4—How can one place the confusing linguistic model at the base of a rigorous mathematical construction?

The proponents of the linguistic model are aware that the use of language runs into some trouble and so put forward some countermeasures in reply to query **Q.4**.

De Finetti explains that the random event should be described in an accurate manner. He adds that we should imitate the insurance companies stipulating contracts under precise and detailed statements. Each particular of the event should be described in order to make clear its unique features.

However, language is fluid to such an extent that words are often to be interpreted. For example, an insurance contract, which de Finetti cites as a perfect linguistic example, is so ambiguous that a magistrate is called to resolve legal disputes about its interpretation. Moreover, in the practice subjectivists accept a description such as the following: "The coin comes down heads". Only two items are mentioned: the coin and the heads; the date, the time, the place and all the particulars, that make this event unique and unrepeatable, remain implicit. In conclusion, there is a noticeable gap between recommendations and professional practice.

Also Carnap is deeply concerned about language and holds that several philosophical problems are indeed pseudo-problems and basically are the outcome of a misuse of language. Some of them can be resolved when we recognize that they are not expressing matters of fact, but they rather regard the choice between different linguistic frameworks. For Carnap, the logical analysis of language becomes the principal instrument to address the probability issues. Since an ordinary language is ambiguous, he asserts the necessity of using artificial languages, which are governed by the rules of logic and mathematics. He proposes to adopt the formal language in order to express the argument of probability with rigor. Mathematicians follow his suggestion when—for instance—they write the probability of x larger than y in this way

$$P(x > y).$$

The probability to see the inter-arrival time X larger than $(t + k)$, given that X is larger than k, is written in this form

$$P[X \geq (t + k)\backslash X \geq k].$$

However, the formal language, which is well grounded in abstract terms, turns out to be rather unmanageable and even harmful in several practical applications. Logicians set forth hypothesis, inferences and conclusions in formal statements, but this linguistic precision cannot be applied everywhere. We miss abundant case studies which could explain the extensive applicability of Carnap's lesson (see also Appendix A).

- All the mathematicians—regardless of the school: frequentist, subjectivist, logician and others—assume the random event as a primitive at the base of their theoretical construction. But the term 'primitive' derives from Latin *'primitivus'* that is the 'first of its kind'. By definition, a primitive is a simple idea or an elementary mathematical entity and can be left to intuition only for the property of being a plane concept. For example, a number and a subset are primitives. The complexity of the linguistic model opposes the assumption of $A = \{p, q, ...\}$ as a primitive; and one could ask:

Q.5—Can we declare the random event complicated and at the same time assume it as a primitive?

In conclusion, Kolmogorov introduces the set model for the probability argument which is subjected to some significant exception. Common misuses and abuses of the linguistic model emerge in a rather evident manner. One could reasonably close that the random event definition constitutes a significant conundrum and it is my opinion that mathematicians should investigate carefully what is the exact object measured by probability and how it should be modeled.

7.3 Precise and Generic Arguments

Honderich (1995) assigns three main senses to the entry '*argument*'. In the most important meaning for philosophers, an argument is a collection of two or more propositions, all but one of which is the premise supposed to provide inferential support for the truth of the remaining one or ones: the conclusion. Secondly, the word 'argument' indicates a quarrel between rather opposite positions. Lastly, one reads: "In mathematics the argument of a function is an input to it, or what it is applied to; while the output, for a given argument, is called the result or value". Computer science offers a similar description and defines 'argument' as a value or reference passed to a function, procedure, subroutine, command or program by the calling program (IBM Dictionary 1994).

Let us dissect the entry in the Oxford dictionary which I rewrite in the following form to make evident that the mathematical concept of argument is two-fold:

In mathematics the argument of a function is

(1) an input to it, or

(2) what it is applied to.

The term 'argument' can be assumed in strict sense and also in broad sense. The first argument is the value processed by the algorithm; the second is the object pertinent to the function. The term 'argument' in strict sense is intended as the input to the calculus algorithm; alternatively, it is the object which the algorithm refers to. For example, take this function

$$y_E = f(x, z, p, q, \ldots). \tag{7.1}$$

The variables $(x, z, p, q, \ldots)$ are used to calculate f, hence they constitute the *precise argument* of (7.1), while the symbol E denotes the *generic argument* of f. By way of illustration, the law of Ohm is obtained from the electrical potential difference V observed across the conductor and the electrical current I through the conductor. The variables V and I are the input to the calculation of resistance and furnish the definition of Ω. The law provides the resistance of the conductor G, and one writes this generic argument as a subscript

$$\Omega_G = \Omega(V, I) = \frac{V}{I}. \tag{7.2}$$

Generally speaking, the rigorous argument of any given measurement justifies the cause/effect relationship occurring in physical reality and provides the definition of the measure. Instead, the generic argument provides subsidiary information. We could call the former as 'definitional' variables, and the latter as 'denotative'. Due to this great difference between the two, the reader perhaps wonders:

Q.6—Can the random event be classified as the precise argument of probability?

Q.7—If not, what is the precise and definitional argument of probability?

In my opinion, the best method to answer queries **Q.6** and **Q.7** consists in inspecting the fundamental equations of the probability calculus. It is better to study mathematics, than to develop philosophical debates.

Masters coming from different theoretical schools base the mathematical concept of probability upon the following statements (Gut 2012).

Mutually exclusive. If the events A and B are incompatible, then the probability of either occurring is

$$P(A \cup B) = P(A) + P(B). \tag{7.3}$$

Independent. If the events A and B are independent, then the joint probability is

$$P(A \cap B) = P(A)P(B). \tag{7.4}$$

Conditional. The probability of A, given B is defined by

$$P(A|B) = \frac{P(A \cap B)}{P(B)}. \tag{7.5}$$

These equations obtain the results by means of probabilities and not through random events. The symbols A and B codify indeterminate facts whose likelihood is being calculated and so one can introduce the formalism adopted in (7.1) without loss of precision

$$P_{A \cup B} = P_A + P_B,$$

$$P_{A \cap B} = P_A \cdot P_B,$$

$$P_{A \backslash B} = \frac{P_{A \cap B}}{P_B}.$$

The symbols A and B denote the objects of which we want to calculate the probability, and do not enter the algorithms of (7.3), (7.4) and (7.5). Thus one can bring the ensuing answer to **Q.6**:

The random event is the generic argument of probability. (7.6)

This conclusion offers some direct advantages. As first, it suggests a reply to:

Q.4—How can one place the confusing linguistic model at the base of a rigorous mathematical construction?

Inaccurate linguistic reports oppose the precision of mathematics and the linguistic model $A = \{p, q, ...\}$ would be incongruent with the calculus of probability on the condition that the random event was the precise argument of probability. Conversely, if the random event is simply the object which the probability refers to, then the incongruence vanishes. One can describe a generic argument using a generic communication mean; accurate statements are unnecessary. The object of the probability measurement may be called on the basis of little information. A symbol used as such in the previous examples is enough, and one can bypass the recommendations of de Finetti and Carnap without detriment.

The present remarks deriving from (7.6) match with universal experience. Languages are equivocal but people normally describe random events using texts, short or long phrases, incomplete or approximate expressions and written or verbal communications. Since time immemorial men and women adopt linguist accounts to figure out chance.

The secondary role described in (7.6) brings forth a reply also to the following question point:

Q.5—Can we declare the random event complicated and at the same time assume it as a primitive?

In mathematics and logic, a primitive notion is usually motivated by an appeal to intuition and everyday experience; but, if the random event is merely the object which the probability refers to, then we can easily mention it and objection **Q.4** falls to the ground.

References

Feynman, R., & Hibbs, A. (1965). *Quantum mechanics and path integrals*. New York: McGraw-Hill.

Gilson G. (2013). Unified Theory of Wave-Particle Duality, the Schrödinger Equations, and Quantum Diffraction. *arAiv:1103.1922v10*.

Green, J. A. (1988). *Sets and groups*. Boca Raton: Chapman and Hall/CRC.

Gut, A. (2012). *Probability: A graduate course*. New York: Springer.

Honderich T. (ed.) (1995). *Oxford Companion to Philosophy*. Oxford University Press.

IBM Dictionary. (1994). *IBM dictionary of computing*. Berkeley: McGraw-Hill Osborne Media.

Kallenberg, O. (2002). *Foundations of modern probability*. New York: Springer.

Kolmogorov A. (1933). *Grundbegriffe der Wahrscheinlichkeitsrechnung*. Springer. Translated as Foundations of the Theory of Probability by Chelsea, (1950).

Vose, D. (2000). *Risk analysis: A quantitative guide*. Chichester: Wiley.

Chapter 8
Structural Modeling of the Probability Argument

The modern calculus of probability proves that the random event is the generic object of ascriptions of probability and the following query waits for an answer:

Q.7—What is the precise and definitional argument of probability?

This problem can be scaled through a progressive method; notably, we could start from the notion of 'event' to arrive at the precise variables that provide the input to the algorithm and furnish the definition of probability.

However we have just seen that the set and the linguistic models have some weak points; they do not depict the event in a perfect manner and some doubts are still in our mind. These models could not support the present discussion and I believe that we should seek for a more accurate representation of the random event in order to implement our project.

This chapter presents the structural model of the event. We shall go back to question **Q.7** in Chap. 9.

8.1 Structural Model

It seems good to me to specify what we usually intend by the term 'event', and this passage by Allison (1984) seems a sharable description:

> An event consists of some qualitative change that occurs at a specific point in time. One would not ordinarily use the term "event" to describe a gradual change in some quantitative variable. The change must consist of a relatively sharp disjunction between what precedes and what follows.

An event can be described as a dynamical change between two extremes that we can generically call as *antecedent* and *consequent* (Fig. 8.1).

P. Rocchi, *Janus-Faced Probability*, DOI: 10.1007/978-3-319-04861-1_8,
© Springer International Publishing Switzerland 2014

Fig. 8.1 Graph model of the event

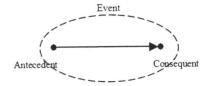

8.1.1 Graph Model

An event is a single episode pertaining to the evolution of the universe and broad literature tackles dynamical phenomena using the *theory of graph* (Diestel 2010). The graph consists of an *arrow* (or *edge*) that depicts the change, and two *nodes* or *extreme points* standing for what precedes and for what follows.

Basically, writers talk about two kinds of physical events. The first one produces the output from the input, and the arrow stands for an operation or action. The second event consists in the transformation of something which evolves from one state to another state. In this case, the arrow stands for the transition of the intended object.

Probability theorists and philosophers are not so inclined to use graphical schemes. They do not like illustrate the random event using a graph, whereas professionals and working statisticians depict indeterminate situations in various contexts with the aid of graphs (Cooke 1991). They draw *decision trees, risk trees, fault tree, regression trees, event trees, influence diagrams* and other forms of graph in decision-making and in other applications (Goodwin et al. 1998; Wahlstrom 1994; Ferry 1988; Harshbarger and Reynolds 2011).

The graph model of the random event can be considered close to the *Bayesian networks* introduced by Judea Pearl (Pearl 2000) and others (Neapolitan 2003). Bayesian networks are directed graphs whose nodes normally represent random variables in the Bayesian sense: they may be observable quantities, latent variables, unknown parameters or hypotheses. Edges stand for conditional dependencies. Basically Bayesian networks are tools specialized to carry on Bayesian analysis.

The wide spreading adoption of graphs in literature and the professional practice should enhance our confidence in this graphical scheme. Diagrams prove to work as effective tools in the analysis of compound events instead this book will use them to disclose the heart of elementary events.

8.1.2 Algebraic Model

Markowitz and Raz (1984) hold that the 'entity-relationship language' can accurately describe all kinds of situations. Hence one can import this formalism in order to assign a precise meaning to the components of the graph by means of this triple

$$A = (\eta_1, \eta_2; \rho). \tag{8.1}$$

Fig. 8.2 Regression tree of prostatic cancer (from Garzotto et al. 2005)

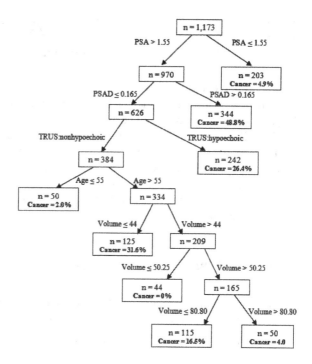

where η_1 and η_2 are *entities* or *sets* of the algebraic space $\mathcal{E}_a$ and ρ is *the directed relationship connecting them*. The elements η and ρ belong to an abstract algebra that is similar to classical algebra except for its use of sets η instead of numbers and the use of relationships to describe operations (Durbin 2009; Tabak 2011) (Fig. 8.2).

Equation (8.1) is a script and (8.2) is a diagram. It may be said that each model depicts the event A from a different perspective. The triple and the graph play complementary roles in favor of the event analysis: the latter is visual, the former is written. One can use the two models in parallel or adopt the most appropriate one to accomplish the project.

$$(8.2)$$

Triple (8.1) and graph (8.2) have different forms and identical content; they do not illustrate the surface but the *inner configuration* or *structure* of events, hence they will be called *structural models* from now onward.

Fig. 8.3 The structural
model stands for the real
event

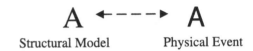

The structural modeling has nothing to do with the current structural studies in probability, which focus on the algebraic–topological aspects of probability theory (Heyer 2004), which reformulate fiducial probability for transformation models (Reid 1992), investigate the error theory and apply other kinds of issues (Fig. 8.3).

The second part of this book aims to explore the argument of probability, in particular and shall examine how the proposed models adhere to reality. We shall look into the practical use of the graph and the triple and we shall argue whether these models are able to represent the phenomenon **A**—certain or indeterminate—which happens in the world.

8.1.3 Use of the Structural Model

The structural model reveals that the interior of an event behaves as a machine whose components act in accordance with precise rules (Johns 2002). In particular:

- The antecedent η_1 is the preliminary element necessary to prompt ρ. In broad sense one could associate η_1 to the so called 'context' of the event.
- The relationship ρ links the antecedent to the consequent and is the lively component of the physical change **A**. When ρ enters into action, we can then say that the overall material event occurs; in substance, ρ allows **A** to happen in the world.
- The consequent η_2 emerges as the output or final state of the event.

The relationship ρ may be defined the *active component* as it runs and establishes a directed connection from η_1 to η_2; the symbol ρ stands for activities, acts, dynamics, developments and/or transitions. The entities η can be seen as *passive components* that symbolize the initial and final states, logical introduction and upshot, inputs and outputs, raw materials, semi-finished products, preliminary components, finished items, outcomes, results. I prefer to quote material situations, but the entities and the relationship can even represent abstract events. For example the antecedent is a certain hypothesis η_1, and the consequent is the evidence η_2. The relationship ρ expresses the logical confirmation of the hypothesis. In this way the structural models fit in with the Carnap approach.

It is not so demanding to detect the entities and the relationships belonging to an event. One can restrict his attention to the parts indispensable for the event to occur; he can pinpoint the essential elements of the change. For example, the roulette can operate thanks to the special device η_o, the hand η_m triggering the movement of the wheel, the little ball ρ_r and the displayed number η_n. These four components constitute the parts of the essence of A_r.

$$A_r = (\eta_o, \eta_m, \eta_n; \rho_r).$$

Table 8.1 Meta-relationships

Name	Symbol
$O\mathcal{R}$	+
$\mathcal{AND}$	•
Conditional relationship	\

The structural model forces one to itemize the indispensable elements which cause changes to occur and bring them into focus. Whereas the natural language conceals misunderstandings, the structural analysis tends to clarify any confusing circumstance. As an example: Will the sun rise tomorrow?

If the probability calculus refers to a possible gravitational collapse, then the structure reports the sun, the earth and their dynamics

$$A_a = \left(\eta_S, \eta_E; \rho_g\right).$$

Instead, if one wonders whether tomorrow he will be able to see the sun rise then the structure encompasses the sun, the obscuring entities η_o (i.e. dense cloud, smoke, vapor, strata, etc.) and personal perception ρ_p

$$A_b = \left(\eta_S, \eta_o; \rho_p\right).$$

The structural model induces to surmount communicative confusion.

8.2 Compound Structures

Fundamental expressions in the probability calculus regard composite events. It is necessary to discuss the primitive parts of (7.3), (7.4) and (7.5) using structural models.

8.2.1 Meta-Relationships

According to current literature, three main relationships link elementary events. Table 8.1 makes the list of the *meta-relationships*: $O\mathcal{R}$, $\mathcal{AND}$ and Conditional relation (Sion 1990).

Let us make a few remarks about the structural analysis of events:

▶ When two or more activities work in turn in the physical reality, they are *disjoined* and one uses $O\mathcal{R}$ to depict this situation. The extraction ρ_1 of a spade card and the extraction ρ_2 of a diamond offer a straightforward example

$$A_{sd} = \left(\rho_1, \rho_2, O\mathcal{R}\right) = \left(\rho_1 + \rho_2\right).$$

One selects the *postfix* or the *infix notation*—illustrated in this case—at his discretion.

▶ When two or more functions operate together in order to arrive at the same scope, they are joined by $\mathcal{AND}$. Cooperating phenomena assume different shapes in physical reality: the four wheels of an automobile turning at the same rate represent an immediate case of conjunction; the king of spades is generated by the parallel extractions ρ_1 of a spade and ρ_2 of a king; a number of spelling errors on the same page represent an example of concurring dynamics against the comprehension of the message; musicians are in $\mathcal{AND}$ since they cooperate for the success of an orchestral performance.

Time and spatial connections can determine cooperation even if cooperating actions are very distant in time and/or in space; i.e. the movement ρ_1 of a cosmic cloud and the run ρ_2 of a star concur to the eclipse of this star which one can be formalized such as $(\rho_1 \mathcal{AND} \rho_2)$. Celestial bodies, distant light-years one from the other, determine this astral conjunction.

▶ Experience shows that an activity can weigh upon another activity or can otherwise modify it in any way; i.e. the extraction ρ_1 of the card E_1 influences the subsequent extraction ρ_2 since E_1 is not put back into the pack and we write

$$(\rho_2 \backslash \rho_1).$$

Conditional operations are very common, and one can find the meta-relationship '\' together with concomitant and alternative ones. The moon and the earth movements concur to the eclipse of the sun; in addition, the movement of the moon ρ_2 depends on our planet and the astronomical conjunction A_S is conditional $\mathcal{AND}$

$$A_S = (\rho_2 \mathcal{AND} \backslash \rho_1).$$

The symbols $\mathcal{OR}$ and\convey somewhat different meanings. The term 'disjoint' means that two or more outcomes cannot be contemporary true. The conditional meta-relationship\signifies that the outcome of one event can influence the outcome of another event. Hence, disjointed events either can be or cannot be independent. To exemplify, suppose the production lines ρ_1, ρ_2 and ρ_3 proceed by turns and bring forth the items E_1, E_2 and E_3 respectively. The element E_3 is an accessory of the product E_2; that is to say, each item E_3 integrates with E_2 and thus the number of the items E_3 carried out in a day-time must be equal to the number of E_2. The overall production A_m is equipped with three disjoint devices while the second machine has a certain influence on the third machine, and the overall structure can be formalized as follow

$$A_m = [\rho_1 \ \mathcal{OR} \ (\rho_3 \ \mathcal{OR} \backslash \rho_2)].$$

The production lines ρ_2 and ρ_3 are mutually exclusive and conditioned at the same time.

Dependence, which is the state of being conditional or contingent on something, is established on a broad inductive basis. It may be said that any physical event is under the weight of something else Ballentine (2001) claims "All probabilities are

Fig. 8.4 Venn's diagrams

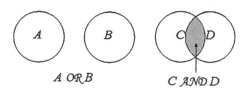

$A \, OR \, B$ $C \, AND \, D$

conditional". It can even occur that an element undergoes two or more treatments at the same time. Two conditional meta-relationships can have different weights and can even act toward opposite directions over the same element. By way of illustration, the worker *wo* supplies the artefacts which he produces to the supervisor *sup*, hence the Markovian dependence yields $(sup\backslash wo)$. At the same time, the worker is subjected to his supervisor, namely $(wo\backslash sup)$. Summing up, there are two reciprocal and opposite effects between *wo* and *sup*

$$A_{ws} = [(sup\backslash wo) \, AND \, (wo\backslash sup)].$$

The first power could be defined as individual and operational; the second one as global and hierarchical. This variable game in mutual influences is very common in economic, commercial, social and political environments. Double conditional effects—similar to contrasting forces—either reach a level of equilibrium or do not reach it. Situations vary from case to case, from time to time.

This just to conclude that conditional relationships are very common and the analysis sometimes becomes somewhat demanding.

Modern writers develop the analysis of an event on the basis of the event's results and visualize the structural configurations by means of the Venn diagrams. For example they depict the mutually exclusive results *A* and *B* with two separated circles. Overlapping circles offer the model of concurrent events *C* and *D* (Fig. 8.4).

It may be said that current literature preferably analyzes the structure η_2 rather than the operations ρ.

8.2.2 Basic Structural Forms

This section is an attempt to show how the graphs and triples turn out to be good tools to analyze intricate events. In general, the meta-relationships create three common configurations called after their graphical forms:

- *Chain.*
- *Tree.*
- *Cycle.*

- A chain is made up of two or more relationships in sequence. It is the most straightforward situation and is well-known in the probabilistic and statistic fields. As the relationship ρ_{k+1} follows ρ_k, thus ρ_k control ρ_{k+1} and one can formalize the Markovian dependence of the first order such as

Fig. 8.5 Fault tree of a car
braking system

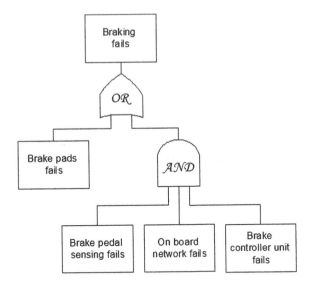

$$\rho = \left(\rho_{k+1}\backslash\rho_k\right).$$

- The relationships ρ_1, ρ_2, ρ_3, ..., ρ_n originating or flowing into the root ρ give a
  tree. For ease, analysts develop the study of a catastrophe through a sequence of
  episodes occurring within a number of given possibilities. The relationships
  pertaining to a tree can run in alternative or in parallel; we find the relationships
  $O\mathcal{R}$, and $\mathcal{AND}$ among the branches of a tree (Fig. 8.5).
- A cycle consists of a repeated action, one that ρ reoccurs as ρ^i, ρ^{ii}, ρ^{iii}, ρ^{iv}, ρ^{v}
  etc. By definition, each operation restarts under the same conditions in a cycle,
  and independence is the fundamental property of this structural form. This
  feature distinguishes the cycle from the chain whose elements are instead
  dependent. Typically, a long-term event creates a cycle and authors specify that
  each trial of the log-term event is independent.

   (8.3)

Relationships ρ^i, ρ^{ii}, ρ^{iii}, ρ^{iv}, ρ^{v}... belonging to a cycle are contiguous in time
or can even occur distant in time and/or in space. For example, we can for-
malize throwing a dice as an enormous cycle whose trials are placed in the five
continents.
Sometimes the operations of a cycle concur towards a unique objective and are
in $\mathcal{AND}$. Take for example:

(1) Six manufacturing processes produce partial results within the global indus-
trial process ρ and pursue the same target

$$\rho = (\rho^i \cdot \rho^{ii} \cdot \rho^{iii} \cdot \rho^{iv} \cdot \rho^{v} \cdot \rho^{vi}).$$

(2) A carpenter creates four table legs and these four operations make up the cycle
ρ to complete the table

$$\rho = (\rho^i \cdot \rho^{ii} \cdot \rho^{iii} \cdot \rho^{iv}).$$

Other applications include the operations of a cycle which are mutually
exclusive. For example, the rounds ρ^i, ρ^{ii}, ρ^{iii},... of a game of chance makes up
a cycle in $O\!R$ when each round occurs now or later, here or there

$$\rho = (\rho^i + \rho^{ii} + \rho^{iii} + \rho^{iv} + \cdots). \tag{8.4}$$

When the cycle in $O\!R$ produces m different kinds of results, thanks to asso-
ciative property we can collect the relationships giving the result η_1; those which
give η_2 and so forth

$$\begin{aligned}
\rho &= (\rho_1^i + \rho_1^{ii} + \cdots) + (\rho_2^i + \rho_2^{ii} + \cdots) + \cdots + (\rho_m^i + \rho_m^{ii} + \cdots) \\
&= (\rho_1 + \rho_2 + \rho_3 + \cdots + \rho_m).
\end{aligned} \tag{8.5}$$

The structural analysis makes evident the difference between the single event
A_1 which comprises one relationship and the long-run event A_n that is a cycle.

8.3 Structure of Levels

A generic event may be described using one or more of the configurations listed
above, and the reader could wonder whether there is a scheme which includes all
the previous structures:
   Is there a general pattern for the events?

A system can be represented in levels of different granularity, and I have
already introduced this technique without any intention. Graph (8.1) exhibits the
event as a whole with the symbol A, and in terms of greater detail as η and ρ.
Triple (8.2) does the same and one can proceed with further steps. One can then
bring to light the sub-entities (η_1, η_2, ..., η_m) and the sub-relations (ρ_1, ρ_2, ..., ρ_p)
and so on in this way

$$A = (\eta; \rho)$$
$$= (\eta_1, \eta_2.., \eta_m;\ \rho_1,\ \rho_2, \cdots \rho_p) \tag{8.6}$$
$$= (\eta_{11}, \eta_{12}\cdots\cdots, \eta_{m1}, \eta_{m2}\cdots;\ \rho_{11}, \rho_{12}\cdots\cdots, \rho_{p1}, \rho_{p2}\cdots)\quad m, p > 0.$$

The following formalism makes explicit the *levels* which are not directly expressed in (8.6)

$$
\begin{array}{ll}
level\,1 & A \\
level\,2 & (\eta;\ \rho) \\
level\,3 & (\eta_1, \eta_{2...}, \eta_m; \rho_1) \\
level\,4 & (\eta_{12}\cdots\cdots, \eta_{m1}, \eta_{m2}\cdots;\ \rho_{11}, \rho_{12}\cdots\cdots, \rho_{p1}, \rho_{p2}\cdots)\quad m, p > 0.
\end{array} \tag{8.7}
$$

One obtains *the structures of levels* (8.6) and (8.7) from the initial structural model in a natural manner.

Expression (8.6) proves to be a more agile script, while (8.7) exhibits the correspondences between the parts and appears to have more explicative qualities. For the sake of simplicity, I assume that the elements of the structure are finite and conclude with the following definition.

The structure of levels S consists of q levels—when q is any non-negative integer—which are indexed sets with this property. If the level j includes Ψ_j and the level $(j + 1)$ includes the elements $\Phi_{j+1}, \Xi_{j+1}, \Theta_{j+1}$ than

$$\Psi_j = \{\Phi_{j+1}, \Xi_{j+1}, \Theta_{j+1}, \ldots\}\qquad j = 0, 1, \ldots, (q-1). \tag{8.8}$$

Each line can include one or more configurations discussed in the previous section. Cycles, chains or trees can describe a level in detail. The structure of levels turns out to be a model flexible to depict a variety of events, as the levels are independent from the nature of the intended phenomena, which can be physical, logical, social and so forth.

The multilevel analysis has infiltrated various technical areas. E.g. the Data Flow Diagram and the Entity Relationship Diagram have been mentioned in the previous pages.

Also thinkers argue over the method of subdividing topics into levels of different granularity. Howard Pattee—a pioneer in the field—in his fundamental article (1973) claims that natural systems appear hierarchical in nature and that science should go forward in a multi-layer study. Philosophers of science and technology are deeply involved in this argument (Ahl and Allen 1996); they investigate the decomposition methods available in ontology (Poli 1998), philosophy of mathematics (Warfield 1986), set theory (Marek et al. 1976), logic (Zadeh 1997), biology (Gross and Lenaerts 2003), management science and other domains of research.

Fig. 8.6 Graph model of the composite event H

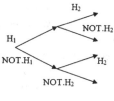

Fig. 8.7 Tree graph of the results of H

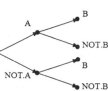

A very broad amount of works illustrates the flexibility of this approach. The subdivision into levels and the graph model are very popular; that is why the present book does not provide several examples.

8.3.1 A Case

As an example, let us apply the structural model to the following case.

> The computer manufacture CyberX sells computer towers (CTs) and video display units (VDUs). CyberX has presented separate bids for CTs and VDUs to a large company. The experience of CyberX in submissions leads it to estimate at 60 % its chances of winning the contract of CTs. If this contract is allocated, the firm assesses its chances to 2 against 1 of obtaining the contract of VDUs. However, if the contract of computer towers is not obtained, the company believes that it still has 30 out of 100 chances to win the contract for the video diplay units.

Let us define the following sub-events:
H_1 Computer tower contract is allocated to CyberX.
H_2 Video display units contract is allocated to CyberX.

While NOT.H_1 and NOT.H_2 are the sub-events H_1 and H_2 that do not occur. The overall compound event includes four alternative situations whose details appear at the bottom of the structure of levels

$$H =$$
$$= (H_{s1} + H_{s2} + H_{s3} + H_{s4}) =$$
$$= (H_1 \cdot \backslash H_2) + (H_1 \cdot \backslash NOT.H_2) + (NOT.H_1 \cdot \backslash H_2) + (NOT.H_1 \cdot \backslash NOT.H_2).$$

The probability that CyberX gets the two contracts is given by the first member in the third level

$$P_{(H1 \cdot \backslash H2)} = (0.60) \cdot (2/3) = 0.40.$$

One can apply the graph model (8.2) to the entire event H in this way (Fig. 8.6).

This fits with the tree graphs currently in use, which however do not derive from a precise model of the random event—as it happens in the present book—secondly practionioners refer the tree graph to the event results. In the following scheme, A and B are the outcomes of H_1 and H_2 respectively (Fig. 8.7).

The probability that CyberX gets the two contracts is calculated in this way

$$P(A \cap B) = (0.60) \cdot (2/3) = 0.40.$$

This case means to show how the structural models offer a precise, consistent support to probability applications.

If we explain the model graph in introductory lessons, then the analysis of the random events would involve the students since the first stages. The exercises in probability should facilitate the professionals-to-be in their future working environments.

Theorists who overlook the random event modeling, involuntary create a gap between the probability foundations and the qualified practice.

References

Ahl, V., & Allen, T. F. H. (1996). *Hierarchy theory: A vision, vocabulary, and epistemology.* New York: Columbia University Press.

Allison, P. (1984). *Event history analysis: Regression for longitudinal event data.* Thousand Oaks: Sage Publications Inc.

Ballentine, L. E. (2001). Interpretation of probability and quantum mechanics. In A. Khrennikov (Ed.), *Foundations of probability and physics* (Vol. 13, pp. 71–85). New York: World Scientific.

Cooke, R. (1991). *Experts in uncertainty.* Oxford: Oxford University Press.

Diestel, R. (2010). *Graph theory.* Heidelberg: Springer.

Durbin, J. R. (2009). *Modern algebra: An introduction.* New York: Wiley.

Ferry, T. S. (1988). *Modern accident investigation and analysis.* New York: Wiley.

Garzotto, M., Beer, T. M., Hudson, R. G., Peters, L., Hsieh, Y. C., Klein, T., Mori, M. (2005). Improved detection of prostate cancer using classification and regression tree analysis. *Journal Clinical Oncology, 23*(9), 4322–4329.

Goodwin, P., & Wright, G. (1998). *Decision analysis for management judgment.* New York: Wiley.

Gross D. & Lenaerts T. (2003). Towards a definition of dynamical hierarchies. In *Proceedings of the Eighth International Conference on Artificial Life* (pp. 45–53).

Harshbarger, R. J., & Reynolds, J. J. (2011). *Mathematical applications for the management, life, and social sciences.* Belmont: Wadsworth Publishing Co Inc.

Heyer, H. (2004). *Structural aspects in the theory of probability.* River Edge: World Scientific.

Johns, R. (2002). *A theory of physical probability.* Chicago: University of Toronto Press.

Marek, W., Srebrny, M., & Zarach, A. (Eds.). (1976). *Set theory and hierarchy theory: A memorial tribute to Andrej Mostowski.* New York: Springer.

Markowitz, V. M., & Raz, Y. (1984). An entity-relationship algebra and its semantic description capabilities. *Journal of Systems and Software, 4*(2–3), 147–162. (special issue).

Neapolitan, R. E. (2003). *Learning bayesian networks.* Englewood Cliffs: Prentice Hall.

Pattee, H. H. (1973). The organization of complex systems. In H. H. Pattee (Ed.), *Hierarchy theory: The challenge of complex systems* (pp. 1–27). New York: Goerge Braziller.

Pearl, J. (2000). *Causality*. Cambridge: Cambridge University Press.

Poli, R. (1998). Levels. *Axiomathes, 1–2*, 197–211.

Reid, N. (1992). Introduction to Fraser (1966) structural probability and a generalization. In: *Breakthroughs in Statistics* (pp. 579–586). New York: Springer Series in Statistics.

Sion, A. (1990). *Future logic: Categorical and conditional deduction and induction of the natural, temporal, extensional, and logical modalities*. Denman Island: Avi Sion Publisher.

Tabak, J. (2011). *Algebra: Sets, symbols, and the language of thought*. New York: Chelsea House Publishers. (Revised edition).

Wahlstrom, B. (1994). Models, modeling and modellers: An application to risk analysis. *European Journal of Operations Research, 75*(3), 477–487.

Warfield, J. N. (1986). Micromathematics and macromathematics. In *Proceedings of IEEE International Conference on Systems, Man and Cybernetics* (Vol. 2, pp. 1127–1131).

Zadeh, L. A. (1997). Toward a theory of fuzzy information granulation and is centrality in human reasoning and fuzzy logic. *Fuzzy Sets and Systems, 90*(2), 111–127.

Chapter 9
Some Topics on Probability and Structural Models

It may be that the preceding chapter provides a misleading vision of the structural analysis to the reader. The simple examples just presented could suggest the idea that an expert is capable of detecting all the elements of a physical event at ease; instead, the contrary is true.

9.1 Demanding Structural Analysis

Let us see some common difficulties encountered by scientists and professionals in the description of events:

- Sometimes an event occurs under inaccessible circumstances or it cannot be observed due to time delays and/or space distances. In other occasions, traces of the facts are cancelled or there is no witness to the truth. Impediments may also be issued by an authority, such as when documents regarding a political fact are covered by State secrecy. The toughest obstacles result in the absolute incomprehension of situations.
- A second ensemble of impediments is comprised of partial and transient ones, such as insufficient financial funds or shortage of resources. Sometimes investigations of reality are possible but only in broad strokes, and one ignores a number of essential details. The reader should also consider inquiries which alter the physical context and consequently lessen the reliability of gained information. Data recording may damage or modify the event itself in some environments. Heisemberg's principle asserts that if we measure an atomic particle, we inevitably change its movement. We can also influence macroscopic items; i.e. destructive biological analyses cause irreversible mutations in human tissues.
- Finally, I remind the reader that people sometimes voluntarily refuse a microscopic account and pay attention only to some essential traits of the event which are easier to hold; i.e. a manager commissions a statistical study and takes his decisions on the strength of global values although he possesses analytical data.

P. Rocchi, *Janus-Faced Probability*, DOI: 10.1007/978-3-319-04861-1_9,

The difficulties of structural analysis match with those philosophers who relate probability to the Man's limited knowledge about the world. From the beginning, thinkers claimed that the probability calculus is an attempt to predict future results using incomplete news, approximated values, and unreliable sources. Human blindness derives from a broad assortment of causes and the calculus of probability is linked in a way to people's ignorance, inability or reluctance to hold a deterministic model. This was the perspective of the authors who developed the calculus of probability from Pascal onward. Boole (1854) describes this view saying that:

> Probability is expectation founded upon partial knowledge. A perfect acquaintance with all the circumstances affecting the occurrence of an event would change expectation into certainty, and leave neither room nor demand for a theory of probabilities.

More recently, Bayesians place the partial knowledge about the state of nature at the center of their methodology. Probability represents the person's uncertainty about propositions when he lacks full awareness of causative circumstances (Idier 2013; Giron and Rios 1980).

Modern manuals illustrate an assortment of cases; writers comment on the generic indications that often lay at the base of risk analysis (Morgan and Henrion 1998), decision making (George 1991) and so forth.

9.2 Illustration of Events

The reader may wonder whether the structure of levels provides assistance to those who have shortage of information:

Does the structure clarify the partial knowledge of events typical of the probability calculus?

One can answer in the following way.

Suppose the structure S stands for the physical event A and has q_S levels ($q_S \geq 0$) (Fig. 9.1). Let us examine the relationship between the 'model' S and the 'original' A; when there is no level

$$q_S = 0. \tag{9.1}$$

The structure is void and in substance people ignore A. If one can merely denote the event, S has only one level and people have generic knowledge of fact without any detail

$$q_S = 1. \tag{9.2}$$

If S includes two or more levels

$$q_S > 1. \tag{9.3}$$

Then there are two alternative cases. Let us examine the structure which can fulfill different duties.

Fig. 9.1 The structural
model stands for the real
event

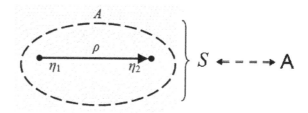

9.2.1 Complete Structure

Suppose the structure $\check{S}$ has q_S levels ($q_S > 1$) and the physical event A has q_A levels when

$$q_S = q_A.$$

This means that relationships and entities of the lower level depict all the typical details of **A**. Further particulars do not appear in the physical reality, and $\check{S}$ provides an exhaustive image of **A**. The structure $\check{S}$ is said *complete*.

Take for example a stone that falls down to earth without attrition loss. The essential elements of the experiment are the initial state η_{S1} of the stone, the final state η_{S2}, the height η_H and the force of gravity ρ_g which moves the stone from η_{S1} to η_{S2}

$$
\begin{aligned}
&\textit{level 1} \qquad A^+ \\
&\textit{level 2} \quad \left(\eta_{S1}, \eta_{S2}, \eta_H; \rho_g \right).
\end{aligned}
\tag{9.4}
$$

In the initial state the stone is at rest and η_{S1} is characterized by null speed. The stone gains the speed v_2 in the state η_{S2} under the gravity acceleration g

$$
\begin{aligned}
\eta_{S1} &= v_1 = 0, \\
\eta_S &> 0, \\
\eta_{S2} &= v_2 = \sqrt{2\eta_H \cdot g} > 0.
\end{aligned}
$$

Structure (9.4) thoroughly depicts the physical event and other significant particulars do not intervene in the world. All the elements included in (9.4) are correct and one can conclude that A^+ occurs for sure.

There is another possibility, in the sense that the structure $\check{S}$ is complete but contains one or more elements which are inexact and do not find precise correspondence in the physical world. As an example take the following structure—similar to (9.4)—which finishes with the state η_{SK} instead of the state η_{S2}

$$\begin{array}{ll} level\ 1 & A^- \\ level\ 2 & \left(\eta_{S1}, \eta_{SK}, \eta_H;\ \rho_g\right). \end{array} \qquad (9.5)$$

where

$$\eta_{S1} = v_1 = 0,$$
$$\eta_S > 0,$$
$$\eta_{SK} = v_K = 0.$$

The stone falls down and cannot have the final speed v_K equal to zero; the event A^- cannot happen in the world.

In conclusion, if the elements of the complete structure Š comply with the elements of A in a perfect manner, then the consequent follows the antecedent and authors say that the event A is *certain*. If the structure Š comprises one or more untruthful elements, then we say that the event is *impossible*.

9.2.2 Incomplete Structure

Suppose

$$q_S < q_A.$$

The model Ş does not exhibit all the details on A. The lower levels of A include details which Ş misses and this structure is said *incomplete*. For example, the structure of the flipped coin that comes up head contains the launching operation ρ, the coin η_c and the outcome η_h

$$\begin{array}{ll} level\ 1 & A_h \\ level\ 2 & \left(\eta_c, \eta_h;\ \rho\right) \\ level\ 3 & \dots \end{array} \qquad (9.6)$$

The interactions of the coin with the fingertips are unintelligible. The elements of level 3, namely the uncountable micro-dynamics of ρ_h, are essential as they produce the final result, even if they do not appear in (9.6). An ideal observer with all the minutia in hand should be able to create the complete model of A and consequently could forecast the result of each experiment. However, he misses out on the particulars and the model Ş is unable to predict the outcome with precision.

The incomplete structure conforms to probabilistic thinking which does not coincide with complete ignorance neither with perfect knowledge. For instance:

- A punter learns the rules of a gambling game. The macroscopic parts of the game are familiar to him, but the minute ones determining the win or the loss remain indistinct.
- The expert of weather prevision has data that describe the atmospheric situation of a region, but misses all the particulars describing the winds, the pressures, the temperature of the soil etc. which are impossible to acquire.
- Individual stories, reflected within each personality, produce the numberless set of human dynamics in communal living and a sociologist resorts to statistical surveys.
- Billions of cells, each of which are made up of hundreds of thousands of chemical components, compose the human body and a doctor is unable to state for sure if his patient will heal.
- Interactions between individuals and groups, or between real and simulated, feared and assured, future and present interests, embody an economic event which nobody can completely dissect.

In conclusion, partial understanding of facts appears to be typical of probability applications and the present theory can but formalize it through the incomplete structure $\mathcal{S}$.

In the literature, there are various positions held about the origin of partial understanding of people. David Hume (1909) is perhaps the most eminent advocate who negates the existence of absolute chance. For Hume, human ignorance leads to all our ideas of probability. The ignorance of the real cause of an event influences the understanding of this event, and begets a sort of belief or opinion with the aid of the probability calculus. More recently, Bayesians hold that uncertainty is an inherent feature of the state of knowledge of the observer, and not the object or phenomenon being observed.

Some quantum physicists have the opposite opinion. The wave function which lies at the core of quantum mechanics is a probability function used to understand the nanoscale world. Using the wave function, physicists can calculate a system's future conduct, but only with a certain probability. Quantum particles behave randomly on their own because that is just what they do. The forecast is fundamentally non-deterministic due to the inherent unpredictability of reality, and is not a limit based on our models or our measurements (Accardi 1990).

In the present logical framework the incomplete structure turns out to be independent from philosophical debates; $\mathcal{S}$ exhibits the available details of the probabilistic event which one grasps regardless the origin of his fractional knowledge and partial ignorance.

9.2.3   Ignorance, Uncertainty and Perfect Knowledge

The present book places apart any philosophical rumination and adopts the analytical approach to the argument of probability, in particular we can sum up the

Table 9.1 Human knowledge of an event

Levels of the physical event	$q_A > 1$	$q_A > 1$	$q_A > 1$	$q_A > 1$
Levels of the structure	$q_S = 0$	$q_S = 1$	$q_S > 1$	$q_S > 1$
			$q_S < q_A$	$q_S = q_A$
Status of the structure	Void	Generic	Incomplete	Complete
Knowledge	Missing	Generic	Uncertain	Perfect

topics just discussed using precise parameters. We set the indeterminate cases discussed from (9.1) to (9.6) into the following ideal scale which relates the structure of levels to the real event A and in turn, to human knowledge capabilities.

The classification (Table 9.1) relies on the dimensions q_S and q_A of the structures and can be defined as an analytical report. At the same time, the spectrum covers the entire range from complete unintelligence to perfect consciousness, and thus the table should be considered an exhaustive overview.

This scheme is not static in the sense that human unawareness can evolve as time went by, and an individual can progress from the ignorance of the event A to its complete knowledge. As matter of fact, researchers pass from preliminary and confused notions to certain ones. Maybe the first statistical survey stops at $q_S < q_A$; later investigations furnish evidence and researchers discover the complete structure with $q_S = q_A$.

9.2.4 Structures and Probability

The complete structure $\check{S}$ is equipped with all the elements necessary to forecast the behavior of an intended event in the world. One has in hand all the elements to predict with precision what will happen or otherwise will not happen. If the elements of $\check{S}$ are correct, the relationship ρ joins together η_1 and η_2, and A modeled by $\check{S}$ surely runs. If the components of $\check{S}$ are wrong, that is to say, η_1, η_2, or ρ are inappropriate, the reported event does not occur in the world. One has a deterministic, precise prediction in both the cases.

The incomplete structure $\S$ does not depict the physical event A in an accurate manner. Significant details on the mechanism ρ are missing. The change does not appear to be subjected to any precise governing design or method, and is called *random event*. As a consequence $\S$ does not allow any precise prevision. The present frame conforms to the idea that the randomness is resulting from the lack of knowledge.

The model $\S$ proves to be insufficient for any precise forecast and all that remains to do is to calculate the proportion of times that the specified event occurs. People normally estimate the number P_A that indicates *how often* A will take place in the world. One can use the structural models to spell out P in this way

$$P_A = P_{(\eta 1, \eta 2, \rho)}.$$

Nobody has the precise model of the situation in hand; in a way A seems mysterious and the probability P_A does not allow people to predict when exactly A will happen. Certain events occur for sure and impossible events are incapable of happening. The probability of indeterminate events is placed between certain and impossible cases

$$0 < P_A < 1.$$

This is just to say that the structure of levels can offer a theoretical reference to the naive concept of probability and correlated ideas such as uncertainty, credibility, indeterminism and randomness.

9.3 In Search of a Precise Argument

Once the structural models of random events have been described, we can tackle the seventh question which is still waiting for an answer:

Q.7—What is the precise and definitional argument of probability?

Publilius Syrus writes: "Practice is the best of all instructors". Following up on this aphorism, one can exploit the concept of probability suggested by universal experience in order to answer Q.7 while the structural model $\mathcal{S}$ offers the theoretical framework of reference.

9.3.1 Experiments

The law of large numbers entails that one can derive the probability from the number of successes detected in a sample of trials, when n is very high.

The sample s has to include trials that have comparable configurations; that is to say, the structure $\mathcal{S}$ of the experiments is equipped with the same antecedent η_1, similar mechanism ρ and a finite number of possible outcomes η_2.

An operator ascertains how often A occurs in physical reality by registering the *relative frequency of A in the sample s*

$$P_A = F_{As} = N_{As}/n. \tag{9.7}$$

Thus the number of success N_{As} and the number of trials n emerge as the precise argument of this expression

$$P_A = \frac{N_{As}}{n} = \frac{number\ of\ experimental\ occurence\ of\ A}{grand\ total\ of\ experimental\ occurence} \cdot \qquad (9.8)$$

The current literature defines (9.8) as *empirical probability* under the constraint that n is much greater than a unit.

9.3.2 Classical Definition

The previous empirical approach inspires the idea that P may be calculated as a quotient even in the abstract form. Fraction (9.8) suggests the following precise criterion to obtain P_A on the theoretical plane.

We normally refer the probability P_A to a set of homogeneous experiments that have the same antecedent, the same relationship and M_A possible upshots, hence one can resolve the structure $\S$ into *ideal cases* or *forms*; that is to say, the abstract event A can be subdivided into the different ways in which A furnishes M_A distinct outputs. Each ideal case $A_{(1)}, A_{(2)}, A_{(3)}, \ldots, A_{(M)}$ emits one of the results which are expected for A

$$\begin{aligned} &level\ 1 \qquad\qquad A \\ &level\ 2 \quad \left(A_{(1)}, A_{(2)}, A_{(3)}, \ldots, A_{(M)} \right). \end{aligned}$$

For example, suppose an ace is drawn from a fair deck of 52 cards. This event comes about in four different forms because four aces are included in the card deck and each ideal case exhibits a different ace

$$\begin{aligned} &level\ 1 \qquad\qquad A_C \\ &level\ 2 \quad \left(A_{(1)}, A_{(2)}, A_{(3)}, A_{(4)} \right). \end{aligned}$$

The number M_A may be defined as *number of theoretical occurrence of A*, normally called *number of favorable cases* in literature.

As empirical probability is calculated in a group of homogeneous events, so the theoretical probability should refer to comparable cases which have the same antecedent, similar relationship ρ while the results belong to a precise set. The entire group of cases includes the favorable cases as a special group. One can assess the grand total of cases m ($m > M_A$) also called *grand total of theoretical occurrences* or *number of possible cases* in literature. For example, the grand total of the ways to draw a card from the deck is 52. The extraction process of each card has identical premise and mechanism, and exhibits 52 different possible results. In fact, one can extract an ace or even another card from the deck.

The number of *theoretical occurrences* or *favorable cases* is consistent with P_A in point of logic, where the larger one is M_A and the more probable is A. Due to the symmetry between P_A and M_A, it is reasonable to assert that the number of favorable cases is proportional to probability

$$\frac{M_A}{P_A} = c.$$

Where c is any positive integer. Then one gets

$$P_A = \frac{M_A}{c} \qquad c > 0. \qquad (9.9)$$

As c can be any, we choose c as being equal to the grand total of theoretical occurrences

$$c = m.$$

Finally, we can assess probabilities in terms of the ratio of favorable cases to all cases

$$P_A = \frac{M_A}{m}. \qquad (9.10)$$

In literature, this equation is often called *classical definition of probability.*

Now consider all the m cases: $A_{(1)}, A_{(2)}, A_{(3)}, \ldots, A_{(m)}$. Each one is a random event and thus we are able to calculate the probability of the generic $A_{(j)}$ ($j = 1, 2, \ldots m$) using (9.10).

We have followed a precise procedure to demonstrate (9.10), in particular we have counted the number of favorable and possible cases. This procedure is intended to intimate that the number of favorable case of $A_{(j)}$ is a unit and thus the probability of the generic ideal case $A_{(j)}$ becomes

$$P_{A(j)} = \frac{1}{m} \qquad j = 1, 2, \ldots m. \qquad (9.11)$$

This expression is true for any j, and this means that result (9.10) is founded on precise assumption (9.11). Equation (9.10) is right on condition that *all the cases are equally likely.* The classical definition of probability is bound to the so called *equiprobability hypothesis* (9.11) because each case is counted singly. This constraint does not meet great difficulties in the working environment. People find that a large number of situations comply with (9.11) and the classical equation operates as a very powerful tool of calculus.

The classical definition which has been illustrated for discrete events is used even with continuous variables. Several significant results have been obtained—see the Buffon's needle problem—in this field.

Fig. 9.2 Classification of
mathematical arguments

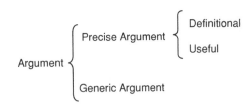

9.3.3 Definitional Arguments of Probability

Equations (9.7) and (9.10) show the precise arguments of probability, and one can now give an answer to question **Q.7**.

On the purely theoretical plane, we get probability from the number of favorable cases and the number of possible cases which can be summarized as *theoretical occurrences*

$$P_A = \frac{M_A}{m} = \frac{number\ of\ theoretical\ occurrence\ of\ A}{grand\ total\ of\ theoretical\ occurrence}.$$

Empirical probability is obtained from this expression

$$P_A = \frac{N_{A_s}}{n} = \frac{number\ of\ experimental\ occurence\ of\ A}{grand\ total\ of\ experimental\ occurence}.$$

One can conclude that:

*The numbers of occurrences---calculated on the paper and detected in
a series of trials---constitute the precise and definitional arguments of* (9.12)
probability.

The numbers of occurrences—theoretical and experimental—spell out how often the random event happens, even if P does not allow people to predict when exactly the event will occur.

9.3.4 Useful Arguments of Probability

We looked over the precise and generic arguments of this mathematical function

$$y_E = f(x, z, p, q, \ldots). \tag{7.1}$$

We restricted our attention to the special form of the precise argument which is able to furnish the definition of the intended output.

Table 9.2 Distribution of blood types

Type	%
A	40
B	9
O	49
AB	2

There is another kind of precise argument that does not define any concept and is used for convenience. This third type of argument could be called '*useful argument*' (Fig. 9.2).

For example, suppose a car is running at a variable speed due to the irregular profile of the road. Velocity is defined as the space change Δs divided by the change in time

$$v = v(s, t) = \frac{\Delta s}{\Delta t}.$$

However, for the driver it is more interesting to understand how the speed changed along the route and he relates v to the traveled distance

$$v = v(d).$$

This function describes the history of the journey from the start to the end, and provides information for the driver who wants to optimize future travels by car. Beyond any doubt, the variable d provides a good and useful service.

Probability applies to broad assortment of issues and experts create probability functions that depend on distance, temperature, volume, and many other parameters. Statisticians adopt either continuous or discrete variables as useful arguments (Table 9.2).

Examples are also found in experiments whose sample space is non-numerical, where the distribution would be a categorical distribution, and in experiments whose sample space is encoded in various ways, etc. By way of illustration, the previous table exhibits the distribution of blood types for a certain population.

Experts are familiar with the *probability density function* (pdf) that can be written in this way

$$P(\Xi \in X).$$

Pdf describes the relative likelihood for the random variable Ξ to take on a given value of the space X. The probability density function is nonnegative everywhere, and—if pdf is continuous—its integral over the entire space is equal to one. If the pdf is discrete, the sum of probabilities from all the possible values of the variable is one.

The *cumulative distribution function* (cdf) $F_\Xi(\xi)$ describes the probability that a real-valued random variable Ξ with a given probability distribution will be found at a value less than or equal to ξ

$$F_\Xi(\xi) = P(\Xi \leq \xi).$$

In the case of a continuous distribution, it gives the area under the probability density function from minus infinity to ξ. A continuous cdf can be defined as the integral of its respective probability distribution function; vice versa pdf of a continuous distribution is defined as the derivative of cdf. Cumulative distribution functions are important in calculating critical values, P-values, power of statistical tests and the distribution of multivariate random variables.

By way of illustration, the following function is the cumulative distribution function of the distance r of the electron in a hydrogen atom from the center of the atom while the distance is measured in Bohr radii

$$P(r) = 1 - \left(2r^2 + 2r + 1\right) e^{-2r}.$$

For example, $P(1) = (1 - 5e^{-2}) \approx 0.32$ means that the electron is within 1 Bohr radius from the center of the atom at 32 % of the time.

References

Accardi, L. (1990). Quantum probability and the foundations of quantum theory. In R. Cooke & D. Costantini (Eds.), *Statistics in science: The foundations of statistical methods in biology, physics and economics*. Dordrecht: Kluwer.

Boole, G. (1854). *An investigation of the laws of thought on which are founded the mathematical theories of logic and probabilities*. London: Macmillan.

George, Ch. (1991). *Decision making under uncertainty: An applied statistics approach*. New York: Praeger Publication.

Giron, F. J., & Rios, S. (1980). Quasi-Bayesian behaviour: A more realistic approach to decision making? *Trabajos de Estadistica y de Investigacion Operativa, 31*(1), 17–38.

Hume, D. (1909). *An enquiry concerning human understanding*. New York: The Harvard Classics.

Idier, J. (2013). *Bayesian approach to inverse problems*. New York: Wiley.

Morgan, M., & Henrion, M. (1998). *Uncertainty: A guide to dealing with uncertainty in quantitative risk and policy analysis*. Cambridge: Cambridge University Press.

Chapter 10
Exploring into the Essence of Events

We have examined how the precise argument $(x, z, p, q, \ldots)$ provides the result of the following function

$$y_E = f(x, z, p, q, \ldots). \tag{7.1}$$

The generic argument plays a secondary role from a mathematical viewpoint but one should not consider the object E as a trivial element since it helps experts to understand the meaning of y. Scientists gain the perfect knowledge of the parameter y when they become aware of the object E measured by y. It may be said that y, $(x, z, p, q, \ldots)$ and E make a conceptual triad; they clarify one another and help theorists to enhance the axioms from which y is derived (Fig. 10.1).

This virtuous circle should encourage probability theorists to pay attention to the argument's issues which I raised in the second part of this book.

At present, the majority of authors refer the concept of probability to the overall random event and/or to a part of it. They use the set model to depict either an event or an upshot while no justification is given of this free choice which contrasts with the precision of mathematics. One could object that this behavior is not so rare. Technicians are allowed to calculate the resistance of an electronic appliance and also of a component of it.

This is true beyond any doubt, but in any case the resistance Ω refers to distinct electrical elements. An engineer does not confuse the resistance of the whole circuit with the resistance of a component. Instead statisticians assign the same value of P to A or to the outcome η_2 with indifference.

In my opinion, we should investigate the most appropriate generic argument which works in synergy with P (Fig. 10.2).

The structural models illustrate the complex texture of A. They demonstrate that an event looks like a machine where the relationship ρ connects the entities η_1 and η_2. One can see the antecedent η_1 as the 'preamble' of the occurrence A and thus appears to be rather nonsensical as argument of P. One wonders whether ρ or η_2 is more suitable than A for the role of generic argument. The reader could ask:

What the proper object of ascriptions of probability is?

P. Rocchi, *Janus-Faced Probability*, DOI: 10.1007/978-3-319-04861-1_10,
© Springer International Publishing Switzerland 2014

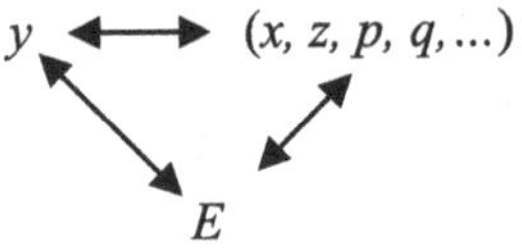

Fig. 10.1

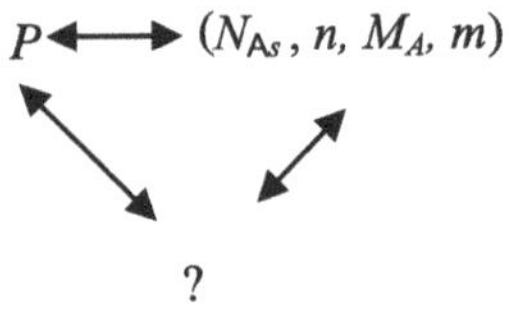

Fig. 10.2

10.1 The Core

The structural models teach us that an event occurs in the world because its dynamical component ρ goes into action. In short, the internal mechanism runs and generates **A**. One can exemplify this concept by quoting the game 'roulette' which is made up of the special device η_o, the hand η_m which starts the spin and launches the little ball, and the displayed number η_n. The spinning ρ_r of the wheel and the ball is the central component of the roulette as it results in the winning number

$$A_r = (\eta_o, \eta_m, \eta_n; \; \rho_r).$$

A change occurs in the world because ρ acts; that is to say, the dynamical component of A runs. The relationship can be reasonably defined as *the core of the event*.

10.1.1 Equivocal Description

Probability theorists do not define ρ in formal terms so far, but authors are aware of the fundamental role played by it. The description of ρ is so important that one cannot calculate the probability if ρ is not strictly fixed. When one does not specify the dynamical component of A, the calculus of P becomes controversial. By the end of the 19th century, Joseph Bertrand (1889) introduced the famous paradox to show that probabilities may not be well defined if the mechanism or method that produces the random variable is not clearly described.

Consider an equilateral triangle inscribed in a circle. Suppose a chord of the circle is placed at random. What is the probability that the chord is longer than a side of the triangle?   (10.1)

Fig. 10.3

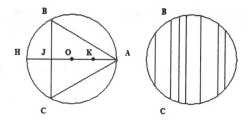

Fig. 10.4

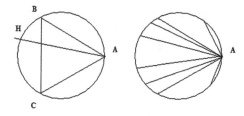

Fig. 10.5

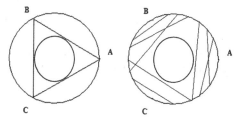

   This brief description conceals different dynamics. As many as six methods of calculus have been identified; I confine myself to three of the best known:

- Let the points H, J, O, K and A equidistant in the diameter. Under this hypothesis the chord proves to be greater than the side of the triangle if it falls within J and K; the chord is smaller than the side when it falls in HJ and in KA. Since any segment is equiprobable, the probability that the chord is longer than the triangle side is 1/2.
- The triangle ABC is equilateral: the arcs AB, BC and CA are equal and the chord has the same probability of having its extreme H in one of the three arcs. The chord is greater than the side of the equilateral triangle, only if H falls on the arc BC and therefore the probability is 1/3.

- When the chord has the mean point inside the inner circle then the chord is smaller than the triangle's side. The probability derived from the ratio of the areas is 1/4.

The terms of problem (10.1) generically regard a chord which is longer than the equilateral triangle side, but this generic property hides the fact that a chord can drop on the circle in different ways. The first method of calculus presumes that the chords fall down randomly and parallely to BC (Fig. 10.3 right). We call ρ_1 this dynamical effect. The second one assumes that the mechanism ρ_2 works in the following manner: one fixes an extreme of the chord in A, while the other extreme H moves randomly along the circumference (Fig. 10.4 right). The last method considers the circle inscribed in the equilateral triangle and ρ_3 causes the chords to fall in between the two circles (Fig. 10.5 right).

In short, the *chord of the circle*—mentioned in (10.1)—can be placed at random in three ways: ρ_1, ρ_2 and ρ_3.

10.1.2 More or Less Steady

The structural model shows that an event is an arrangement of working parts very similar to a car whose motor engine looks like ρ. This interpretation enables us to look into the event from the operational viewpoint. In particular, one can observe that the event A becomes real as soon as the relationship ρ starts to run; ρ may be considered the mover of A or even the component which causes A to be brought into action. Formally one can conclude that A is true if and only if ρ is true

$$\rho \Leftrightarrow A. \tag{10.2}$$

The relationship causes the physical occurrences to take place in the world; and the more often ρ operates, the more often A comes into being in reality. By definition, probability says how often an event occurs, thus on the basis of (10.2) one can adopt the relationship as a generic argument

$$P = P\rho.$$

The generic argument ρ makes evident that P measures something that is dynamical. Probability regards occurrences that are mental actions and physical operations as well; P quantifies a fact that takes place in abstract or in the real world.

One can go on with the operational analysis which casts light on the properties of ρ.

By definition, the relationship establishes a link between η_1 and η_2; as such it may be more or less stable. The relationship joins two entities and this work may be more or less subjected to fluctuations. A practical example can clarify what I mean to say.

Fig. 10.6 The rope tying two balls

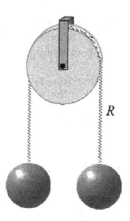

The rope R links two heavy balls placed at the two extremes. Suppose one has r pairs where each is made up of two balls that have more or less the same weight; that is to say, the second pair weighs approximately the same as the first one; the third pair is roughly the same, and so on. Suppose one uses R to sustain the first couple of balls; later on R sustains the second couple, and so forth. There are three possibilities (Fig. 10.6):

- The cord R sustains r pairs of balls with success.
- The cord is incapable of sustaining all the pairs because it gets broken every time (we suppose the rope can be restored in a perfect manner whenever it splits apart).
- The cord sustains k pairs of balls ($k < r$) and breaks down with ($r - k$) pairs.

The rope R is a good metaphor of the relationship ρ which has variable capability of joining. This operational property can be related to the notion of probability and the following cases exemplify the quality of ρ with regard to P:

1. All living beings are destined to die; that is to say, death ρ_D changes a living body η_L into a dead body η_D. Given the antecedent, the relationship ρ_D operates without doubt because the demise is certain for any living being even if no-one can predict the actual date.
2. Suppose the event consists of a rolled fair dice which—when it comes to rest—shows the number seven on its upper surface. The outcome seven is impossible and the value $P\rho_7 = 0$ tells that never and ever ρ_7 results in η_7.
3. Suppose the rolled dice η_C comes to rest and shows the number four. The rolling mechanism ρ_4 is neither sure nor impossible. The probability $P\rho_4 = 1/6$ explains that ρ_4 connects the input to the output from time to time without a governing design.

From the operational point of view it may be said that the connection ρ is more or less strong, and consequently the success of ρ turns out to be more or less frequent. Suppose that processes ρ_A and ρ_B have probabilities $P\rho_A = 0.1$ and $P\rho_B = 0.2$; it may be said that the connection ρ_B is twice as solid as ρ_A.

The operational qualities of ρ can be summarized as follow:

> *The relationship ρ connects two entities and $P\rho$ quantifies this*   (10.3)
> *connection which may be more or less steady.*

Property (10.3) is independent of the nature of the event which may be anything at all: social, biological, logical, and so forth. The stability of ρ turns out to be a criterion appropriate for various environments; even for abstract issues. As an example, the antecedent is a certain hypothesis H and the consequent is the evidence E. The conditional probability $P_{H\backslash E}$ fixes the confirmation of the hypothesis because the conditional relationship—more or less fluctuating—is placed between H and E. The more stable the relationship \, the higher will be $P_{H\backslash E}$.

The generic argument ρ offers other advantages; it can justify some properties of probability which currently we consider as intuitively evident. The next sections illustrate two results that one obtains using operational criteria.

10.1.3 Multiplication Law

Suppose that the event A includes the sub-events $A_{(1)}$, $A_{(2)}$, $A_{(3)}$, ..., $A_{(r)}$ which pursue the same purpose, that is to say, the sub-events are in $\mathcal{AND}$

$$\begin{array}{ll} \text{level 1} & A \\ \text{level 2} & \left(A_{(1)} \cdot A_{(2)} \cdot A_{(3)} \cdot \ldots \cdot A_{(r)}\right). \end{array} \qquad (10.4)$$

For example, the probability of getting the ace of spades A_{AS} can be computed as the probability of two concurrent events: drawing an ace A_A and drawing a spade A_S

$$\begin{array}{ll} \text{level 1} & A_{AS} \\ \text{level 2} & \left(A_A \cdot A_S\right). \end{array}$$

The sub-events co-operate to reach the same target. We have seen how ρ is the mover of the event, hence one can conclude that also the relationships of the structure (10.4) are in $\mathcal{AND}$

$$\rho = \left(\rho_1 \cdot \rho_2 \cdot \rho_3 \cdot \ldots \cdot \rho_r\right).$$

Due to the variable performance of ρ expressed by (10.3), it is reasonable to assume that the more steady the generic sub-relationship ρ_k, the more stable the overall connection

$$P\rho \text{ is a increasing function of } P\rho_k \qquad k = 1, 2, \ldots r. \qquad (10.5)$$

Generally speaking, the cooperation of a restricted number of elements is preferable to reach an objective since the more numerous are the contributions requested for the same target, the more difficult is the overall conjunction of the parts. Here one can assume that the higher is r, the more difficult it becomes to coordinate ρ_1, ρ_2, ..., ρ_r and lead them to a positive result

$$P\rho \text{ is a monotonic decreasing function of } r. \tag{10.6}$$

From assumptions (10.5) and (10.6), and from P that varies from 0 and 1, it follows that the probability function is multiplication

$$P\rho = P_{(\rho 1 \cdot \rho 2 \cdot \rho 3 \cdot \ldots \cdot \rho r)} =$$
$$= (P\rho_1 \cdot P\rho_2 \cdot P\rho_3 \cdot \ldots \cdot P\rho_r). \tag{10.7}$$

The verification of this result is a straightforward exercise.

Result (10.7) is in line with the formula that calculates two independent events when they both happen together.

10.1.4 Addition Law

Now suppose the sub-events $A_{(1)}$, $A_{(2)}$, $A_{(3)}$, ..., $A_{(r)}$ be mutually exclusive

$$\begin{array}{ll} level\ 1 & A \\ level\ 2 & \left(A_{(1)} + A_{(2)} + A_{(3)} + \cdots + A_{(r)}\right). \end{array} \tag{10.8}$$

For example, suppose to draw the aces from a card deck. This event includes four extractions in $O\mathcal{R}$ since the aces are four

$$\begin{array}{ll} level\ 1 & A_{AS} \\ level\ 2 & (A_1 + A_2 + A_3 + A_4). \end{array}$$

The relationship ρ is the working component of A, and also the sub-relationships of the structure (10.8) are in $O\mathcal{R}$

$$\rho = (\rho_1 + \rho_2 + \rho_3 + \cdots + \rho_r).$$

Property (10.3) is true here too, and one assumes (10.5) once again

$$P\rho \text{ is a increasing function of } P\rho_k \qquad k = 1, 2, \ldots r. \tag{10.9}$$

Operational criteria teach us that a large number of components which we can substitute anyone with another one, pledges the overall result. One can conclude that the more the elements may be replaced in turn, the more the overall mechanism is reliable. In other terms, the greater r is, the more ρ is capable of functioning

$$P\rho \text{ is a monotonic increasing function of } r. \tag{10.10}$$

Under the constraint that the overall result has to be greater than zero and lower than one, and from assumptions (10.9) and (10.10), one can deduce that the probability function is addition

$$
\begin{aligned}
P_\rho &= P_{(\rho 1 + \rho 2 + \rho 3 + \cdots + \rho r)} = \\
&= (P\rho_1 + P\rho_2 + P\rho_3 + \cdots + P\rho_r).
\end{aligned}
\tag{10.11}
$$

This result matches with the Kolmogorov axiom that defines mutually exclusive events.

10.2 About Event's Outcome

The relationship ρ makes the existence of the random events clearer on the theoretical plane. One can deduce multiplication and addition laws from $P\rho$, and in consequence one could reduce the number of axioms. When I discovered these charming features, I became so enthusiastic about them that I wrote a book in order to sustain the new perspective opened by the concept of relationship (Rocchi 2003).

During the compilation of the book, I did not think carefully about the use of $P\rho$ which—see Chap. 9—meets heavy obstacles in practice. The indeterministic structure $\S$ lacks the minute details of ρ that cannot be easily detected. People cannot recognize the occurrences of a random event because of this defect.

By contrast, the outcome η_2 proves to be a clearly visible signal. The relationship is impossible to discern, whereas the result appears evident especially in gambles that emphasize this difference; i.e. a roulette player cannot identify ρ_{16} from the movements of the ball and the wheel tracking device; instead he can easily recognize the number 16.

Experimentalists systematically ascertain the occurrence of an event using its output; the concepts of event and upshot overlap even in theoretical frames. The outcome proves to be a good marker of A and one can write

$$
\eta_2 \Rightarrow A.
\tag{10.12}
$$

Hence, one may assume the result as an appropriate generic argument of probability

$$
P = P\eta_2.
$$

Results are as clear as they reveal the structure of composite dynamics and usually one calculates the basic equations of probability using the set models. One obtains independent probability from two separated sets

$$
P_{\eta_X \cup \eta_Y} = P_{\eta_X} + P_{\eta_Y}.
\tag{10.13}
$$

He gets joint probability on the basis of the set intersection

$$P_{\eta_X \cap \eta_Y} = P_{\eta_X} \cdot P_{\eta_Y}. \qquad (10.14)$$

Experts recognize the structure of compound events from the analysis of the results that factually derive from the operations

$$\begin{aligned} \eta_X \cup \eta_Y &\Rightarrow \rho_X + \rho_Y, \\ \eta_X \cap \eta_Y &\Rightarrow \rho_X \cdot \rho_Y. \end{aligned} \qquad (10.15)$$

However, the problems are not over. A physical event does not always emit an upshot. Secondly some results appear equivocal; in addition the results of some events cannot be modeled as a set (various cases have been commented on in Chap. 7). The outcome η_2 presents some limitations in the role of generic argument and statements (10.12) and (10.15) cannot be considered absolute truths.

Three questions raised around this topic are waiting an answer:

Q.1   How can we justify the set model for the entire event and for a part of it?

Experts calculate P_A or alternatively P_{η_2} at will; this careless switching does not raise serious issues in applications because both A and η_2 are generic arguments of P. Obstacles arise at the theoretical level and the remaining queries pinpoint this sore side of the probability theory:

Q.2   How can we calculate the probability value if the set does not appear in the world or cannot identify the random event?
Q.3   Why does the set model fail?

In conclusion, the overall event A, the relationship ρ and the result η_2 have virtues and vices; they have strong and weak points when we use them as generic arguments of probability. The puzzling enigma remains.

Writers have not formally clarified the object of the probability calculus. This is not a dramatic failure in applications but proves to be a severe shortage on the abstract plane as the axiomatization of probability appears controversial as long as one does not say what he relates the probability to.

Perhaps we should do something in order to ameliorate this cultural situation.

10.3 Final Remark

This book exhibits two different expressive styles and contents.

The first part confronts the problem of probability interpretation and puts forward a precise solution which consists of two theorems. The discussion of those results involves some considerations on the methodology of science with the main aim to reconcile frequentist and subjectivist conceptions of probability.

The second part tends to stimulate a reflection regarding the axioms of the probability theory. The analysis of the probability argument is proposed as the staple topic of discussion and as the obliged passage toward the probability axiomatization. We have examined the generic and the precise arguments of probability with the help of structural models. The former argument does not seem disputable, whereas the latter raises some issues and the object to which probability refers to does not appear clearly modeled until now.

The first part of the book assumes a somewhat polemic stance versus current theories and the second part does not furnish any definitive solution. Regardless of the different contents and diverging purposes, a common thread unifies the pages of this book which rejects any philosophical mode and systematically dissects the minutest parts of the topics under examination. I imagine that everybody recognizes the merits of philosophers who argued over intricate questions, who anticipated some scientific ideas and provided a supporting shelf for elevated culture; although the definitive solution to scientific issues cannot be established through philosophical disputes. I believe that we can clarify the foundations of probability using the analytical approach and it is this approach which is the unifying leitmotif of the present book.

References

Bertrand, J. (1889). *Calcul des Probabilités*. Paris: Gauthier–Villars.
Rocchi, P. (2003). *The structural theory of probability*. New York: Kluwer/Plenum.

Appendix A
Interpretations of Probability—
An Outline

Classical

Pierre Simon Laplace (1749–1827) was one of the most eminent French mathematicians in the 19th century who left his distinctive mark in various areas of physics and astronomy. He was still a young in 1773 when he announced his earlier results in the probability calculus during lectures later published as memoirs (Laplace 1904).

In 1795 Laplace held a course on probability at the École Normal, and later on printed the lessons of his course as 'Essai philosophique sur les probabilités' (Laplace 1814). This book hosts an assortment of topics besides the principles of probability theory; the author argued about the games of chance, natural philosophy, moral sciences, testimony, judicial decisions and mortality.

Laplace's great treatise on probability appeared in 1812 with the title: 'Théorie Analytique des Probabilités'. This deals with an ample set of mathematical topics which include generating functions, approximations to various expressions occurring in probability theory, Laplace's definition of probability, conditional probability, least squares, Buffon's needle problem, inverse probability, with applications to mortality, life expectancy, duration of marriages, and legal matters.

Later editions of the 'Théorie Analytique' contained supplements which consider applications of probability to errors in observations; the determination of the masses of Jupiter, Saturn and Uranus; triangulation methods in surveying; and problems of geodesy. This spectrum of contents gives an idea of the author's prismatic intellectual activity, leading to a new level of mathematical foundation and development to probability theory and mathematical statistics.

***

The second book of the 'Théorie Analytique' begins with the title 'General principles of the probability calculus' where Laplace establishes *the definition of probability as the ratio of the number of favourable cases to the whole number of possible cases* under the assumptions that they are equally possible. This statement emerged from a long series of slow processes which dragged on for several centuries. Laplace takes on an idea which was clearly 'in the air' since a millennia. Pioneer writers from Geronimo Cardano to Bernoulli, from Pascal to De Moivre

P. Rocchi, *Janus-Faced Probability*, DOI: 10.1007/978-3-319-04861-1,
© Springer International Publishing Switzerland 2014

divided the favourable cases by the total possibilities. Sound practical judgment guided Laplace who recognized the origin of his inspiration in this way:

> On voit, par cet Essai, que la théorie des probabilités n'est, au fond, que le bon sens réduit au calcul; elle fait apprécier avec exactitude ce que les esprits justes sentent par une sorte d'instinct, sans qu'ils puissent souvent s'en rendre compte (Laplace 1825). (One sees, from this Essay, that the theory of probabilities is basically just common sense reduced to calculus; it makes one appreciate with exactness that which intellectuals feel with a sort of instinct, often without being able to account for it.)

When one has the probabilities of some simple events, Laplace explains how to determine the probabilities of compound events. He calls the probabilities of the simple events *possibilities* which may be determined in the following three ways:

1. *A priori*: one sees, from the nature of the events, that they are possible in a given ratio; for example, in tossing a coin, if the coin is homogeneous and its two faces are entirely alike, we judge that heads and tails are equally possible.
2. *A posteriori*: one repeats the experiment which produces the event in question and sees how often the event occurs.
3. *By considering the reasons*: one is allowed to pronounce on the existence of the event; for example, if the skills of two players A and B are unknown, since we have no reason to suppose that A is stronger than B, we conclude that the probability of A winning a game is 1/2.

The first method gives the *absolute possibility* of an event; the second one gives the *approximate possibility* and the third gives only the *possibility relative to our knowledge*.

***

It may be said that Laplace's work dominated the scientific field until the early 20th century to the extent that his probability definition was defined by the epithet *classical*. However, the mounting adoption of statistics in various scientific sectors and working environments brought the theory of Laplace into a progressive crisis. The main objection against Laplace can be divided in two claims:

- The theory is circular.
- The theory is inconsistent.

Laplace establishes ten principles to fund his theory of probability. The first one is the very definition of probability; the second principle says:

> Mais cela suppose les divers cas également possibles. S'ils ne le sont pas, on déterminera d'abord leurs possibilités respectives dont la juste appréciation est un des points les plus délicats de la théorie des hasards. Alors la probabilité sera la somme des possibilités de chaque cas favorable (Laplace 1825).

The term "également possibles (equally possible)" simply means "equally probable" so by defining equal probability in terms of equal possibility the theory runs into a vicious circle.

Laplace tends to classify equally possible outcomes by means of the *principle of non-sufficient reason* which dated back to Jakob Bernoulli and perhaps alluded to Liebniz' *principle of sufficient reason*. In brief, this principle states that if we are ignorant of the ways an event can occur, and therefore have no reason to believe that one way will occur preferentially compared to another, the event will occur equally likely in any way. This statement appeared controversial to several authors in particular Keynes renames this criterion as *principle of indifference* and devotes an entire chapter of his book in an attempt to refute it (Keynes 1921).

Various commentators have devised situations where different ways of phrasing what the initial set of equally possible cases are, will lead to different and incompatible probability distributions.

Keynes proposes this case: Suppose we have an urn, with three balls—white and/or black—in it: What is the probability that there are three black balls in the urn?

One application of the principle will be as follows: for each ball, there are two possible cases either it is white or it is black. Altogether, there are eight equally possible configurations of the balls in the urn. According to the principle then, the possibility of three black balls is 1/8.

Richard von Mises censures Laplace's theory and notices some practical cases where one cannot apply the Laplace definition. For example, suppose that the probability of getting head from a loaded coin is 0.67. In this game and in all the games whose conditions are not ideal, how can we discern the success cases and the possible ones?

As a second example, suppose that the probability of death of a Californian born in 1905 and working in the publishing industry is 0.021. What method allows us to analyze the probability? Are perhaps the favorable cases 21 and the total cases 1000; or they are 42 and 2000; or even 63 and 3000?

Von Mises concludes that the classical definition of probability applies to well structured situations and unfits the most difficult real cases, such as the probabilities of life insurance.

Frequentist

The interpretation proposed by P.S. Laplace was widely accepted throughout the nineteenth century and the early twentieth century. His solution was regarded as the only legitimate definition of probability. Although, some thinkers—such as A. de Morgan and G. Boole—began to pinpoint his weak points and presented alternative interpretations immune to any vicious circle. Leslie R. Ellis (1817–1859) and John Venn (1834–1923) outlined a frequentist thesis. Antoine Augustin Cournot (1801–1877), another precursor of the frequentist school, carried out accurate studies into social events and in 1843 he collected his findings in 'Esposition de la Théorie des Chances'. In this work Bernoulli's theorem took a fundamental position and established a steady continuity between theoretical calculus of probability and the practice.

These attempts, however, were not successful in the sense that they did not modify the situation within the scientific community in a significant manner. Richard von Mises (1883–1953) has the merit of having conducted the criticism of Laplace's definition in a vigorous manner (Heyde et al. 2001). From 1919 onward, von Mises (1963/1964) published a series of articles in which he pointed out significant flaws of the classical theory and outlined a new conceptualization which he completed with 'Probability, Statistics, and Truth' (von Mises 1928).

Moreover, von Mises provided outstanding contributions to various subjects of research such as aerodynamics, mechanics, statistics, and positivistic philosophy.

***

Von Mises addresses the probability definition in a very clear and drastic manner as he links the concept of probability to the notion of '*kollektiv*' which he drags from the terminology of sociologists and psychologists. He says:

> It is possible to speak about probabilities only in reference to a properly defined collective.

For von Mises, a collective is an infinite sequence of trials of an event whose possible outcomes have each a definite probability but otherwise appear entirely at random. Abstractly the collective may be represented by an infinite sequence of points of an appropriate space, or by an infinite sequence of natural numbers.

Von Mises defines the mathematical probability as *the limit value of the relative frequency in a collective*. He establishes two fundamental principles in order to sustain this statement:

(i) The relative frequency of a particular attribute within the collective tends to a fixed limit; in other words, it is assumed the *global regularity* of the collective.

(ii) A fixed limit is not affected by any selection; any part of the sequence has the same fixed limit of the infinite sequence. This means that the collective has *local irregularity*.

For von Mises, probability can be controlled by means of measures, and is something independent from man. Probability is a physical quality that people can come to know by studying long sequences of events which contain arbitrary elements (von Mises 1951). This objective existence of probability suggested the appellation '*objective interpretation*' to the studies of von Mises:

> Probability is a science of the same kind as geometry and theoretical mechanics (...) which aims to reproduce observable phenomena not just as a reproduction of reality but rather as its abstraction and idealization (von Mises 1963/1964).

Von Mises' position towards the foundation of probability is intimately connected to positivistic philosophy; he does not esteem that statistical explanations in science are of transient utility while deterministic theories are the definite goal. In fact positivists trust in the universal value of scientific knowledge, regardless the deterministic or indeterministic form of knowledge. This conviction sustained von Mises who spent a certain amount of energy in the statistical science.

The works of von Mises brought about the abandonment of the classical definition of probability on the part of statisticians and his influence on sharpening the opposing standpoints stimulated original studies on the fundamentals of the probability calculus. Mathematicians and philosophers continued to debate throughout the remainder of the twentieth century.

***

A first objection against the philosophy of von Mises regards the notion of limit. In fact, this concept applies—strictly speaking—only to infinite series. Hans Reichenbach (1891–1953) aims at reformulating the frequentist theory of probability in order to revamp its inner logic. He claims that the frequency theory of probability is reducible to deductive inferences with the addition of induction by enumeration. Since there is no proof in the case of empirical series that the limit of the series exist, Reichenbach suggests to posit that in the future the same relative frequency will hold. In his mind the verb '*to posit*' is something like '*to bet*' and '*to wager*'. One does not know whether there is such a limit, but if the limit exists, the Reichenbach's method will find it in the long run (Reichenbach 1935).

Another point slurred over by von Mises is that one can only estimate, but never know, a probability value. Nobody can calculate the relative frequency of events repeated *ad infinitum* and, therefore, one obtains a distorted representation of the very probability. For example, even if a perfectly symmetrical coin is available, no finite sequence of trials could do more than suggest that the probability for heads is in the neighborhood of 1/2. The gap between this estimate and the conclusion that "*P* is exactly 1/2" cannot be easily filled.

Dissent arises against the consistency of a collective because of the unclear definition of the randomness postulate (ii). The actual existence of a collective cannot be proven in a strict mathematical sense, since the requirement of the irregularity of the collective precludes mathematical rules for its construction. Wald and A. Church propose two accurate criteria to specify the selection of a collective but those criteria restrict the original notion of randomness in a way as notes J. Ville, E. Borel, M. Fréchet and others.

Creating a collective in the real world provokes further difficulties (Ville 1939) For example, tossing a coin for an infinite number of times raises the following question: To be considered a collective, how similar must the tosses be? If the tosses are identical, then the outcome will not change. If they are dissimilar, how much dissimilarity is allowed?

Although the collective concept was first embraced by physicists, it was subsequently rejected by others like Bohr and Schrodinger, both of whom were influenced by Heisenberg's principle of uncertainty. This principle defined uncertainty and probability without the collective concept.

This sequel of objections is due to thinkers who basically accepted the frequentist philosophy and could be defined as *internal criticism*. Non-frequentist authors, which create *external criticism*, raise more severe dissent. De Finetti,

Ramsey and Bayesians place the blame on how the frequentist definition regards exclusively long-run events and is narrow in applications. Von Mises' definition only applies to inherently repeatable events. The relative frequency definition of probability is too confining whereas any mathematical construction should be adopted wherever it is required.

A detailed account of all the objections that had been raised against von Mises' philosophy was given by Fréchet (1938) at the Geneva conference. The reader can find the more recent exhaustive contributions of Hájek (1996, 2009).

Axiomatic

Andrei Nikolaevich Kolmogorov (1903–1987) was one of the foremost mathematicians in the twentieth century. He furnished a variety of important results in probability calculus such as the Kolmogorov–Smirnov test for probability distributions (Kolmogorov 1933a), the Chapman–Kolmogorov equation and the algorithmic theory of complexity (Kolmogorov 1968). However there is no doubt that the most famous and influential work of Kolmogorov (1933b) is a monograph of around 70 pages published in 1933. This monograph changed the character of the calculus of probability, moving it from a collection of calculations into a consistent mathematical construction.

Commentators frequently relate this work to a pair of antecedents.

In the beginning of the 20th century, David Hilbert presented a list of twenty-three ineludible problems in mathematics at the International Congress of Mathematicians, held in Paris. The sixth problem deals with the "mathematical treatment of the axioms of physics" and mentions the axiomatic foundations of probability. Hilbert wrote:

> The investigations on the foundations of geometry suggest the problem: to treat in the same manner, by means of axioms, those physical sciences in which mathematics plays an important role; in the first rank are the theory of probabilities and mechanics. As to the axioms of the theory of probabilities, it seems to me desirable that their logical investigation should be accompanied by a rigorous and satisfactory development of the method of mean values in mathematical physics, and in particular in the kinetic theory of gases (Gray 2000).

Hilbert himself devoted much of his research to the sixth problem; in particular he worked on the axiomatic basis of quantum mechanics. The success of *Hilbert space* methods ushered in a very fruitful era for functional analysis.

In 1898 Émile Borel published an article on the measurability for subsets and for this reason he is usually considered the founder of the mathematical theory of measure (Hawkins 2001). Soon after Henri Lebesgue developed a significant theory of measure on the real numbers $\mathbb{R}$, on the n-dimensional space $\mathbb{R}^n$, and a theory of the so-called Lebesgue-integral in this space. In 1913 Radon unified Lebesgue and Stieltjes integration by generalizing countably additive set functions. Two years later, Maurice René Fréchet extended Radon's theory to any space so long as the countably additive set function is defined on a σ-field of its subsets.

***

Kolmogorov accomplished Hilbert's guideline in the sense that he pointed out some axioms to fund the probability calculus and exploited the measure theory elaborated by Fréchet. The advent of measure theory at the turn of the 20th century gave him appropriate tools to formulate a robust model of probability; in particular the 1930 Radon–Nikodym theorem became a crucial factor in Kolomogorov's exposition (Wang and Klir 2009).

The axioms fixed by Kolmogorov can be rewritten in modern terminology as follows:

Let $\mathcal{F}$ be a set of subsets (called *events*) of Ω:

- $\mathcal{F}$ is σ-algebra of sets.
- $\Omega \in \mathcal{F}$.
- P is a function from $\mathcal{F}$ to $[0, \infty)$.
- $P(\Omega) = 1$.
- If $A \cap B = \varnothing$ then $P(A \cup B) = P(A) + P(B)$.

He adds a sixth axiom—called *continuity axiom*—which is useful when the set of event is infinitely countable:

If $A_1 \supset A_2 \supset \ldots$ is a sequence of events in $\mathcal{F}$ such that $P(\cap_j A_j) = 0$, it follows that:

$$\lim_{j \to \infty} P(A_j) = 0.$$

Today the triple $(\Omega, \mathcal{F}, P)$ is called *probability space*; the set Ω—also called *space of elementary events*—has no structure and represents the set of all possible outcomes of the random experience. A probability space is a measure space with total mass equal to one and the random variable is a real-valued measurable function. Kolmogorov shows that the system of axioms i–vi is consistent, that is to say no statement in it can be both true and false; but adds:

> Our system of axioms is not, however, complete, for in various problems in the theory of probability different fields of probability have to be examined (Kolmogorov 1933a).

The previous axiomatization permits to eliminate the ambiguity caused by various paradoxes in the calculus of probability. The strength of Kolmogorov's theory lies in the use of a totally abstract framework, i.e. the space of elementary events Ω is not equipped with any topological structure. This does not imply that in some particular problems, like the convergence laws, it is convenient to work on better spaces through the use of image measures.

Kolmogorov concerned himself with connecting the theoretic probability with the empirical reality and devotes significant efforts to spelling out the links between the world of experience and the abstract theory. He believes that probability should be interpreted in terms of frequencies, and to avoid the vicious circle he presents two principles (Hendricks et al. 2001). The former is a version of von Mises's postulate that probabilities should be observed in frequencies. The latter principle is a strong form of Cournot's principle. Suppose the random event E happens or does not happen in experiment C, the principles can be expressed in the following terms:

A. One can be practically certain that if C is repeated a large number of times, the relative frequency of E will differ very slightly from $P(E)$.
B. If $P(E)$ is very small, one can be practically certain that when C is carried out only once, the event E will not occur at all.

The *extension theorem* is another solution contrived to give good sense to the appropriate application of the axiomatic theory. The theorem states that a suitably 'consistent' collection of finite-dimensional distributions can define a stochastic process; in other words, a consistent family of finite dimensional distributions is sufficient to completely specify a random process possessing the distributions. This result provides realistic requirements for the construction of mathematical models of physical phenomena.

***

The Kolmogorov formalization has been progressively adopted. Virtually all current mathematical work on probability uses this measure-theoretic approach nonetheless some criticism has been raised against this axiomatization, especially by de Finetti who writes

> Events are restricted to being merely a subclass (technically a σ-ring with some further conditions) of the class of all subsets of the space (in order to make σ-additivity possible but without any real reason that could justify saying to one set 'you are an event' and to another 'you are not'; people are led to extend the set of events in a fictitious manner (i.e. not corresponding to any meaningful interpretation) in order to preserve the appearance of σ-additivity even when it does not hold (in the meaningful field) rather than abandoning it (De Finetti 1970).

Quantum physicists discovered that the two slits experiment does not match with the definition of conditional probability given by Kolmogorov and Accardi (1982) supposes that in reality this definition constitutes a 'hidden' postulate. Authors frequently compare the Kolmogorov probability to the Euclidean geometry and Accardi helps himself with the following similitude. For centuries, theorists believed that the definition of two parallel lines as 'never meeting lines' was a self-evident and rather compulsory notion. Instead one can conceive a geometry where two parallel lines meet in a point. Accardi concludes that in a similar manner the definition of conditional probability is a non-obvious concept and can be regarded as sort of hidden axiom which Kolmogorov omitted to declare.

Lastly, someone deems the Kolmogorov's approach is an evasion to the problem of interpretation. Formal obeisance to the idealization implicit in mathematical theory construction serves merely as an excuse for shirking the hard work of articulating the links between theory and practical applications.

Subjectivist

In the mathematicians' community, the idea that probability can express a personal belief penetrated for a long time. I quote Cournot (1843) who openly talked about 'subjective probabilité'. Other pioneers reverted to this idea in generic terms that is why Savage considers the 'Treatise on Probability' (1921) published by John Maynard Keynes (1883–1946) as "the earliest account of the modern concept of personal probability".

***

Frank Plumpton Ramsey (1903–1930) presents the concept of subjective probability in 'Truth and Probability' (1931). He begins with critical comments on the works of von Mises and Keynes. Ramsey relates that the latter recognizes the subjectivity of probability but in substance does not assign any value to subjectivism. Moreover Keynes believes there is an objective relationship between knowledge and probability, as knowledge is disembodied and not personal. By contrast Ramsey analyses the connection between the subjective degree of belief an individual has in a proposition and the probability it can be given.

Ramsey searches for a behavioural way of measuring the degrees of belief, and answers this question by means of the old established way, that is simply by proposing a bet, and by seeing "what are the lowest odds which he will accept". The strategy is to offer the agent a bet on the truth value of the proposition p involved in the belief. Ramsey considers this method to be 'fundamentally sound' but concludes that:

> It suffers from "being insufficiently general, and from being necessarily inexact (…) partly because of the diminishing marginal utility of money, partly because the person may have a special eagerness or reluctance to bet".

Thus Ramsey puts forward a refined betting method with differences in utilities rather than with money. In this way he circumvents some obstacles opposed by the concept of bet which is not sufficiently general; it varies with marginal utility of money and is influenced by risk. Ramsey seeks to separate probability from utility by the device of an *ethically neutral proposition*. In this way, he obtains *probabilities that are interpretable as measures of pure belief*, uncontaminated by marginal utilities for money.

The *representation theorem*—a cornerstone of his edifice—states that a subject's preferences can be represented by an utility function determined up to a positive linear transformation (Bradley 2004) The representation guarantees the existence of a probability function related to the desires and beliefs of an agent. In this way Ramsey laid out an original approach to subjective probability through the concept of expected utility.

Modern literature accredits Ramsey and Bruno de Finetti (1906–1985) as the founding fathers of the modern school of subjectivist probability. In the same year they published a work on the subjective probability, independently from each other.

De Finetti (1931a) shared the same interpretation of probability and nothing else, since they were very different in many respects. Frank Ramsey was an authentic polyhedron; he made remarkable contributions to epistemology, semantics, logic, philosophy of science, mathematics and metaphysics; Bruno de Finetti was basically a statistician. The latter had a long life which he entirely devoted to his theoretical discovery. His production was abundant. The former died young and his principal work in probability was published after his death. De Finetti had a few followers in his country during his lifetime; Ramsey was highly regarded by the British statistical school—in particular at Cambridge University—which was rich in key figures. I mention Richard Braithwaite who introduced Ramsey to Keynes and posthumously edited Ramsey's papers in the form of a book. W. E. Johnson and Jeffreys also interacted with Keynes and others at Cambridge. Turing, Good, Barnard, and Lindley all fit within the legacy of this Cambridge tradition.

De Finetti essentially proposed the same behaviorist definition of probability without being aware of Ramsey's work, but gave a different foundation to his logical frame, and presented several original cues. First, it may be said that for de Finetti the bet model constitutes a convenient application for talking about probability in a way that makes it understandable to a layman. De Finetti adopts other ways of measuring probability by means of scoring rules based on penalties.

Second, de Finetti does not discuss the value of money and assumes that its utility is linear. It may be said that for de Finetti a bet occurs *hinc et nunc* (here and now), so his probability is effectively the price that he is willing to pay.

De Finetti was an ardent subjectivist; the aphorism: "Probability does not exist" epitomizes his position which denies there are 'correct' probability assignments. He takes a radical approach also when he establishes that all the assumptions of an inference ought to be interpreted as an overall assignment of initial probabilities. The shift from prior to posterior probabilities has a subjective value, in the sense that going from prior to posterior assessments involves a shift from one subjective probability to another. Since the probability absolutely relies on the individuals' preference, de Finetti feels the need of fixing a rule of coherence as a necessary condition for degrees of belief being rational. He selects the Dutch Book criterion and introduces both the concept of coherence and the condition of invulnerability as a necessary and sufficient condition.

De Finetti points out:

> Every probability evaluation essentially depends on two components: (1) the objective component, consisting of the evidence of known data and facts; and (2) the subjective component, consisting of the opinion concerning unknown facts based on known evidence (De Finetti 1970).

By this, he intends to say that probability reflects an individual's credence and, in addition, this credence is also 'objective' in the sense of being operationally measurable, for example by means of betting behavior or scoring rules.

De Finetti combines the subjective notion of probability in terms of coherent beliefs with that of *exchangeability* (de Finetti 1937). In so doing, he is able to guarantee the applicability of subjective probability to practical situations, including those encountered within experimental science. De Finetti calls *exchangeable* those sequences where the places of successes do not make a difference in probability. Exchangeability expresses a type of ignorance: no additional information is available to distinguish among sequences of results. The condition of exchangeability for a sequence of results is stronger than the assumption of identical marginal distributions, and is a weaker condition than independent and identically distributed assumptions (de Finetti 1931b).

De Finetti provides a precise result—he himself calls it '*the fundamental theorem of probability*' (de Finetti 1970)—which gives applicability to subjective probability by bridging degrees of belief and observed frequencies. According to this theorem, exchangeable beliefs over infinite sequences of observable Bernoulli quantities can be represented as mixtures of independent trials. Thus, coherent previsions enable an individual to act as if he first assigns probabilities to events and then determines the marginal prices he is willing to pay for a bet on the basis of the expected values. For subjectivist authors, this theorem generates a sequel of consequences such as tight connections between subjectivist and frequentist reasoning, the proof of priors' existence, an interpretation of parameters which differs from that usually considered, a solution to Hume's problem of induction and other outcomes.

***

A number of severe attacks have been conducted against the subjective interpretation of probability in the initial periods. Von Mises and other frequentists do not recognize any scientific value to subjective probabilities, and even hold that the probability of a single event is nonsensical. For instance, the phrase 'probability of death', when it refers to a single person, has no meaning at all for a frequentist. One cannot say anything about the probability of death of an individual even if one knows his condition of life and health in detail.

Another sequel of blames revolves around the arbitrariness of the subjective determination of the probability value. The theory has no provision to ensure that individuals with identical background information will declare identical probabilities. By way of illustration, suppose an encounter between two strangers in which one reveals that his true probability of an event is 0.1 and the other reveals that his true probability for the same event is 0.2. What should either of them do with this information? Should they revise their beliefs? Should they bet against each other?

Moreover one could question: Observing the two strangers, what should a third party advise them to do?

The question of whether and how the probabilistic beliefs of different individuals should be combined is one of the most vexing problems in the subjective theory. Given an individual's action, it is difficult to separate the individual's probabilities from his or her utilities. Subjectivist authors claim to determine a unique measure of pure belief from observations, but in fact, the uniqueness of the belief measure depends on the choice of a particular small world and on the arbitrary convention of scaling the cardinal utilities identically in all states. In other words, the uniqueness of the belief measure depends on various elements that go beyond the theoretical accounts.

Bayesian

Thomas Bayes was an English Presbyterian minister and an excellent amateur mathematician. His paper: 'An Essay towards Solving a Problem in the Doctrine of Chances' reversed what was the usual focus of reasoning: from the population to the sample. The paper put forth a number of mathematical propositions and presented what is now known as *Bayes' theorem.*

Bayes never published his discovery, but his friend Richard Price found it among his notes after Bayes' death in 1761, re-edited it, and published it. Bayes' theorem was even presented to the Royal Society of London but laid cold until the arrival of Laplace who gave a much more elaborate version of the inference problem and adopted a more modern language and notation (Laplace 1774).

In the first half of the 20th century Bayes' law became to draw increasing attention of specialists and after the Second World War Bayesianism literarily exploded. This scientific movement has been so ample and extensive in episodes as historians of the statistical science find difficult to pin point its evolution and the major actors of its progress (Fienberg 2006).

As regards the earlier part of this period, we restrict attention to Abraham Wald (1902–1950) and others who ameliorated the techniques in statistical decision theory and introduced the label 'Bayes' throughout their work (Wald 1939). Later on, the term *'classical statistics'* came into use to describe frequentist as opposed to Bayesian methods.

The Second World War created significant, even dramatic needs to the nations in war, and stimulated intensive, focused research in various fields. In particular there was a renewed interest in statistical decision theory, and this led to a melding of developments surrounding the role of subjective probability and new statistical tools for scientific inference. This kind of studies flourished in England and U.S. in a special manner. Alan Turing who guided the team specialized in cryptanalysis at Bletchley Park, was also involved in the estimate of the probability of a hypothesis, allowing for the prior probability, when information arrives piecemeal. Turing devised a Bayesian approach to sequential data analysis, using the strength of evidence. In U.S. the Manhattan Project stirred up various studies in the statistical field and led the development of the Monte Carlo method which subsequently provided a significant aid to Bayesians' computation.

Amongst the most prominent personalities, it is worth mentioning Leonard Jimmie Savage (1917–1971) who was active in the field since the thirties. He discovered the works of Bruno de Finetti in the forties and entered in contact with him. During the war he cooperated with John von Neumann and in the fifties sustained the Bayesian approach with energy. Savage met various researchers, such as the British Dennid Lindley and involved him in the development of the Bayesian methodology (Lindley 1965).

Savage (1954) published 'Foundations of Statistics' in which he put forward the theory of personal probability. In a way, Savage synthesizes the works of Ramsey and de Finetti and also incorporates features of von Neumann and Morgenstern's expected utility theory. Recognizing that the marginal utility of money could vary across states of the world and levels of wealth, Ramsey and Savage present a measurement scheme for utility and then tie their definitions of probability to bets in which the payoffs are effectively measured in utiles rather than dollars. In particular, Savage introduces the notion of a prize whose utility would be, by definition, the same in every state of the world. Savage traces the strands underlying Bayesian statistics and develops applications to game theory.

The story of Robert Schlaifer (1914–1994) at the Harvard Business School appears absolutely original. He conceived the basic ideas of statistical decision theory independently of the work of Savage and in addition was able to clarify some noteworthy aspects for application. Schlaifer (1961) wrote 'Applied Statistical Decision Theory' with Howard Raiffa and introduced the idea of conjugate prior distributions. Raiffa (born 1924) provided significant contributions in game theory, behavioral decision theory, risk analysis, and negotiation analysis.

The Bayesian movement is as rich of ideas as it resembles a tree with several branches. No question whether the various branches of research go toward rather diverging directions.

The 'objectivist' Bayesian statistics as opposed to 'subjectivist' Bayesian statistics searches for optimal objective procedures in light of a well-defined scenario. According to the objectivist view, the probability can be seen as an extension of the logic definition of probability. Harold Jeffreys (1891–1989) and Edwin Thompson Jaynes (1922–1998) are perhaps the most known advocates of this strand of studies (Jeffreys 1961), (Jaynes 1986).

Ronald Aylmer Fisher (1890–1962) introduced a method to arrive at a posterior degree of belief without mentioning a prior. His *fiducial probability* could be defined as a 'heretical' doctrine respect to 'orthodox' Bayesian doctrine. Fisher's theory attracted noticeable controversy and was never widely accepted.

***

The Bayesian interpretation of probability starts from the Bayes theorem which prompts experts to learn from experience and data. The Bayesian definition of probability begins from the inverse probability approach of updating degrees of belief in propositions. *The probability is defined as a quantity that one assigns theoretically, for the purpose of representing a state of knowledge, or that he worked out on the base of previously assigned probabilities.* By observing the outcome of the first trial, one realizes the second probabilistic value. The first trial provides an individual with some insight into the system, and it may be that his probability for the second trial will change from what it is now. One repeats several trials and continues to update his personal belief.

The historical excursus—above outlined—should be sufficient to comprehend that Bayesians did not stop on the probability definition. They did not want a *descriptive theory*; instead they worked for a *normative theory* which could be useful in the working environment: the Bayesian statistics. Experts have derived a large set of statistical tools and procedures, which progressively have gained space.

The statistical methods start with existing 'prior' beliefs, and update these using data to give 'posterior' beliefs, which may be used as the basis for inferential decisions. Bayesian logic—obviously I intend the 'orthodox' logic—allows an expert to express conditional logic statements and statistical techniques are often used for logic reasons. They allow an expert to make a conclusion about the likelihood based on what he knows, i.e. the best answer.

***

For decades Bayesians have been accused of a number of defects that often are not real failures but are misinterpretations depending on the completely different philosophy of Bayesianism respect to the frequentist framework. Bayesians are fully aware that common terminology is not well adapted to comprehend their methods. For instance, the Bayesian analysis treats θ as though it was a random variable whereas classical analysis treats θ as a fixed constant, albeit unknown. And the notional truth behind the binomial sampling model, is that θ is fixed, not random.

Besides special and rather technical problems, the fundamental opposition against the Bayesian definition of probability relies on the personal nature of probability. Frequentists hold that any scientific measure should be grounded with objective facts, such as the frequencies of outcomes obtained in repetitions of similar experiments. In Bayesianism, any probability is a conditional probability given what one knows but that could vary from one person to another. The inferred plausibility depends on whether one takes the past experience of one person, of a group, or of humanity. A probability derived from an individual's personal judgment is severely charged by the followers of exact sciences; whilst experts in social sciences such as medicine, economy and sociology do not find so obscene the Bayesian personalism.

The reactions of Bayesians to theoretical attacks cover a broad range. On the one hand, we find the Bayesians sensitive to the accusation of arbitrary. For instance, the 'objectivist Bayesian' theory searches for considerable extrinsic justification as a method of analysis. On the other hand, there are experts absolutely indifferent to the previous criticism and derive the extreme consequences from the subjectivist philosophy. The studies on *imprecise probability* and imprecise decision theory generalize the fundamental theorem of de Finetti and handle credence that do not need to be exact numbers (Walley 1991).

Logical

During the Middle Ages thinkers distinguished the *formal* or *minor logic* from the *material* or *major logic*. The former had to develop and verify correct reasoning apart from its connections with the material world; the latter not only had to fulfill this duty but in addition had to verify the results in the world, notably it had to establish the 'truth' of science. Progressively experts abandoned this second vein of research and restricted their attention to the formal logic which actually became synonymous with 'abstract' logic.

Gottfried Wilhelm Leibniz (1646–1716) intervened in the discussion upon the mathematical treatment of logic and distinguished the *deterministic* proofs which lead to certainty, and the *probabilistic* demonstrations (Leibniz 1989). It is not clear what Leibniz envisioned and what would be a means for *estimating likelihoods*; anyway the new logic emerging in the second half of the 19th century was created in the Leibnizian spirit.

George Boole (1815–1864) argued about the close relationship between the logic relations *not, and,* and *or* and the formal properties of probability in a book which after the Second World War became famous as theoretical basis of the digital technologies (Boole 1854). The central problem for Boole was to obtain the probability of an event in terms of the probabilities of other events logically related to it in any manner. He expresses the event as a proposition and in the second part of his book develops a method which is based on his logical calculus developed in the earlier part of the book.

Subsequently, Augustus De Morgan (1806–1871), Charles Sanders Peirce (1839–1914), Ludwig Wittgenstein (1889–1951) matured contributions tending to clarify the connection between logic and probability. In particular, it may be said that in Wittgenstein's thinking about probability there are two poles: the first pole is the logical theory of probability as it is sketched in the 'Tractatus'. The second pole is the epistemological view of probability as it is briefly outlined in 'Philosophical Remarks' and 'Philosophical Grammar'. Wittgenstein theory of probability is linked to the notions of imperfect knowledge and incomplete descriptions. He says that the logic of probability is only concerned with the state of expectation in the sense in which logic is concerned with thinking.

It is widely recognized that Keynes and Rudolf Carnap (1891–1970) played a significant role in the development of modern inductive logic.

In the 'Treatise on Probability' the first extended the deductive logic of conclusive inference to an inductive logic of inconclusive inference by postulating a relation of 'partial entailment', knowable a priori. This relation enables a probability measure of the strength of an inference from one proposition to another. But in 'Truth and Probability' Ramsey attacked the idea of an a priori inductive logic so effectively that Keynes abandoned it. Despite Ramsey's rejection of Keynes' ideas on logical probabilities, they were later taken up and relaunched by Carnap.

Carnap was a philosopher who investigated into various topics such as the structure of scientific theories, the inductive and modal logic, and the philosophy of language; he was also an advocate of logical positivism. He worked on explicating inductive probability from the early 1940s until his death in 1970. He made the basis of his theoretical proposal in 'Logical Foundations of Probability'. His last and perhaps best contribution was published posthumously in two parts, in 1971 and 1980, with the title 'A Basic System of Inductive Logic'.

Carnap inherited the central elements of his probability interpretation from Keynes. Unlike Keynes, there was a deeper analysis of the formal language in Carnap's researches. He agreed on the concept of *logical syntax* proposed by Wittgenstein in the 'Tractatus' and formulated and implemented the details of logical syntax to the extent that Carnap suggested to substitute logical syntax for metaphysics.

One can see a certain evolution in Carnap's view and in the formal techniques which he used. In the last period of his production there was a shift from sentences in a formal language to subsets of a sample space. This reflected in part the desire to get closer to the formalism of common probability calculus. Secondly, he accepted the De Finetti-Ramsey-Savage link of probability to utility and decision making and this placed Carnap squarely in the Bayesian camp. We have to recall the positive significant contribution given by Carnap to the Bayesian methods.

∗∗∗

Carnap gives an account of the probability as a degree of confirmation which can be popularized in the following terms. Suppose an individual intends to examine a certain hypothesis H. He makes many observations of particular events which he regards as relevant for judging his hypothesis. He formulates this evidence and the results of all observations made in the report E. Thus, he tries to decide whether and to what degree the hypothesis H is confirmed by the evidence E, and the *measure of this confirmation is the probability* $P(H|E)$. As an example, H is that a coin will land heads up, E is that a fair coin is about to be tossed; the plausible confirmation is $P(H|E) = 1/2$. Carnap points out that E entails H only partially but this partial entailment can be established in a precise and objective manner using a formal language which expresses H and E in rigorous terms; hence the probability can be defined as *the probability of the linguistic proposition H given the proposition E.*

The most significant consequences of the present approach arises from the assertion that the probability of a statement, with respect to a given body of evidence, is a *logical relation* between the statement H and the evidence E. Thus, it is necessary to build an *inductive logic* construction that studies the logical relations between statements and evidence; it may be said that inductive logic provides the mathematical method of evaluating the reliability of a hypothesis. One can evaluate the degree of confirmation of H and, in principle, one could compare alternative theories.

Inductive logic fits with the continuous scale of probability values and at the same time is two-value logic. In fact, this dichotomy is not between truth and falsity of a sentence but between implication and not implication for two sentences. The probability $P(H|E) = 1/2$; is either true or false and does not have an intermediate truth-value of $1/2$.

***

There are many criticisms of Carnap's inductive logic. A group of objections address the formal-technical aspects of his theory such as the consistency of axioms (Costantini and Mondadori 1973). Others censure the 'total evidence' which is not well defined and the overvaluation of the syntactic rules. Notably he was not able to formulate a theory of the inductive confirmation of scientific laws; he did not discuss the procedural limitations of inductive logic.

Other objections focus on practical aspects such as the scarce applications of the theory which misses clarifying critical topics. Carnap developed somewhat simple cases that are not applicable to situations of real interest; he overlooked continuous variables, infinite sets and situations with rich background.

The robust and varied dissent of thinkers has been gathered by Hájek (2003a) who offers a useful compendium of the literature.

Propensity

In my opinion, one cannot discuss Karl Popper's (1902–1994) thought on probability independently from the philosophy of science developed by him.

It is widely accepted that science is characterized by the empirical method, essentially inductive which proceeds from observations to conceptualization. Popper sees that also some pseudo-science adopts the empirical-inductive method in a way, thus the central problem for him becomes how to distinguish genuine science from pseudo-science (Popper 1963). He devotes his best energy to specify the features of the former and the latter, that is the *problem of demarcation* between science and non-science. It is not to solve a problem of meaningfulness or significance, nor truth, nor acceptability; it is neither a question of explanatory power for the Austrian philosopher. Popper confronts the problem of demarcation in a very original manner. He claims that a theory should be considered scientific if, and only if, it is falsifiable through experimental tests. Every test of a genuine theory is an attempt to falsify it, or to refute it. A theory which is not refutable by any test is non-scientific; radicalizing, irrefutability is not a virtue of a theory but a vice.

The criterion for the scientific status of a theory is its testability but there are various degrees of undergoing a test: some theories are more testable, more exposed to refutation, than others.

Popper talks about the use of tests and closes with a historical conclusion: no number of positive outcomes at the level of experimental testing can confirm a scientific theory. There is no direct epistemic support that a theoretical hypothesis can gain via observation. Confirmation of this is a myth. By contrast, a single counterexample is logically decisive to deny a theory. The truth content of a theory, even the best of them, cannot be verified by scientific testing, but can only be falsified. The term *'falsifiable'* does not mean something is made false, but rather that, if it is false, it can be shown by observation or experiment.

When a theory turns out to be incompatible with certain results of test, the theory is to be refused. Alternatively, a genuine theory, when found to be false, can still be upheld by introducing ad hoc auxiliary assumptions, or by re-interpreting the theory ad hoc in such a way that it escapes refutation. Sometimes such a procedure can be followed, but it rescues the theory from refutation at the price of confining or at least narrowing its scientific field of action.

Popper (1994) holds that the more information a statement contains, the greater will be the number of ways in which it may turn out to be false; hence, the higher the informative content of a theory the lower will be its probability. Informative content is inversely proportional to probability and directly proportional to testability. A statement with a high probability will be scientifically uninteresting, because it says little and has no explanatory power; by contrast, an interesting and powerful statement has low probability. Hence Popper declares his great concern in improbable but powerful theories. Due to the central role played by testability and falsifiability, Popper concludes that he is interested in theories with a high degree of corroboration, even if those theories appear highly improbable. In this way, the Austrian philosopher challenges positivism, which held that one should prefer the theory most likely to be true. This quarrel is only a little sample of the extensive involvement of Popper in the theory of probability which crossed his entire lifetime.

***

Earlier on in his career, Karl Popper (1935) adhered to the frequency interpretation of the probability, but later he expanded his inquiries in the field and began to discuss other interpretations of the term 'probability', such as the logical interpretation and the subjectivist interpretation. In particular, after the Second World War Popper (1957) paid increasing attention to single-case probability motivated by phenomena occurring in quantum mechanics (QM).

He definitively rejected both the frequentist and the subjectivist reading of probability, and instead of somehow reconciling them, he made a third proposal that is different from both: *the propensity theory* (Popper 1959). In addition to explaining the emergence of stable relative frequencies, the idea of propensity is motivated by the desire to make sense of single-case probability attributions.

This topic involved him for a long while and the elaboration of Popper does not appear linear in the sense that he modified his position while works progressed. Popper's evolving thought emerges from his conspicuous philosophical production. Notes, appendix and corrections in manuscripts and in published works testify his laborious and creative concern.

In search of the constant traits in the rich elaboration of the propensity concept, we find two conceptual cornerstones:

(a) The strong links between probability and physical phenomena.
(b) The intent of providing the comprehensive account of probability.

(a) Since the beginning, the Austrian philosopher saw probability as something related to the physical world and this idea progressively led him toward the notion of propensity.

Propensities are probabilistic dispositions very similar to material dispositions and Popper explains himself by quoting a comparison case. He reminds that dispositional properties are named by words that have '-*ble*' suffixes i.e. a solu*ble* substance dissolves when immersed in the appropriate way. In parallel, random events possess a special sort of probabilistic disposition, called *propensity*. The notion of probability has its origin in a disposition, or tendency of a given type of material situation to yield an outcome of a certain kind, or to yield a long run relative frequency of such an outcome. Popper sees the propensity $P(A)$ as an objective property of A within the context in which A happens. They are the physical properties of the dice—such as the cube faces are smooth, the edges are sharp, the cube falls in a plane. These explain the fact that the limit of the relative frequency of a number is 1/6.

(b) Several authors tacitly accepted the multiple views on probability in the ancient past. More recently eminent writers—i.e. Cournot and Ramsey—declared their pluralist conviction but did not develop a theoretical frame. They are no more than 'tolerant' thinkers. Others seem to begin a pluralist theory, such as Carnap (1945) who ends up placing the different interpretations of probability under the umbrella of inductive logic.

Popper was the first—at the best of our knowledge—to make a methodical attempt to examine the various views from an independent perspective and to place these views inside a unified theoretical frame. He has the merit of having observed carefully the probability issues with an open mind. In search of an intellectual synthesis, Popper (1959) tends to set the notion of propensity into two major boxes. Propensity is a conjectured or estimated statistical frequency in long (actual or virtual) sequences. This is usually called as "the probability of an event of a certain kind" such as obtaining a six with a fair die. Second, the propensity refers to a singular event such as the probability of obtaining a six in the ninth throw made after ten o'clock this morning with this die. One can ask of these propensities whether they are long-run or single-case-properties. Popper himself was undecided on this issue, but whatever the answer is, it has nothing to do with bringing frequentism and subjectivism into agreement. Conversely, it appears evident his original search of a 'third way'.

***

The general idea of propensity is the subject of ongoing discussion. Poppers and his followers are accused of giving a somewhat empty account of probability (Hájek 2003b) (Hitchcock 2002). Others such as Humphreys (1985) indicate logical incongruencies; Shanks (1993) underlines the incompatibility of the propensity theory with relativist physics. Eagle (2004) raises severe objections to the variants of propensity theories which fail to give an adequate explication of probability, or else fail to provide better contents than other conceptualizations.

I believe that the authentic weak point of the Popperian construction is that he had no enough time to complete his project. Outcomes remained unfinished and the various applications of the propensity concept appear rather apart from each other in his production, and irreconcilable in many respects.

The tenet of propensity bifurcates toward different directions in the subsequent literature; i.e. see the classic, single-case and causal dispositional versions of the propensity interpretation. There are propensities that bear an intimate connection to relative frequencies and there are causal propensities that behave rather differently (Gillies 2000). The assorted versions of the propensity theory have been developed by Popperians who underline either an aspect or another one. For instance, Fetzer (1983) emphasizes single-case propensity theories while Hacking (1965) suggests a long-run propensity theory.

References

Accardi, L. (1982). Some trends and problems in quantum probability. In *Proceedings of 2nd Conference on Quantum Probability and Applications to the Quantum Theory of Irreversible Processes* (pp. 1–19). Berlin: Springer.

Boole, G. (1854). *An investigation of the laws of thought on which are founded the mathematical theories of logic and probabilities*. London: Macmillan.

Bradley, R. (2004). Ramsey's representation theorem. *Dialectica, 58*(4), 483–497.

Carnap, R. (1945). The two concepts of probability: the problem of probability. *Philosophy and Phenomenological Research, 5*(4), 513–532.

Carnap, R. (1962a). *Logical foundations of probability*. Chicago: University of Chicago Press.

Carnap, R. (1962/1980). A basic system of inductive logic. In R. C. Jeffrey (Ed.), *Studies in inductive logic* (Vol. 1–2). Berkeley, CA: University of California Press.

Costantini, D., & Mondadori M. (1973). Induzione e Probabilità. *Le Scienze, 58*, 48–55.

Cournot, A. (1843). *Exposition de la Théorie des Chances et des Probabilités*. Paris: Hachette.

de Finetti, B. (1931a). Sul Significato Soggettivo della Probabilità. *Fondamenta Matematicae, 17*, 298–329.

de Finetti, B. (1931b). Funzione Caratteristica di un Fenomeno Aleatorio. *Atti della R. Academia Nazionale dei Lincei, Serie 6, 4*, 251–299.

de Finetti, B. (1937). La Prévision: Ses Lois Logiques, Ses Sources Subjectives. *Annales de l'Institut Henri Poincaré, 7*, 1–68.

de Finetti, B. (1970). *Teoria della Probabilità*. Torino: Einaudi (Trans. *Theory of Probability*. New York: Wiley (1974).).

Eagle, A. (2004). Twenty-one arguments against propensity analyses of probability. *Erkenntnis, 60*(3), 371–416.

Fetzer, J. H. (1983). Probability and objectivity in deterministic and indeterministic situations. *Synthese, 57*, 367–386.

Fienberg, S. E. (2006). When did Bayesian inference become 'Bayesian'? *Bayesian Analysis, 1*(1), 1–40.

Fréchet, M. (Ed.). (1938). *Exposé et Discussion de Quelques Recherches Récentes sur les Fondements du Calcul des Probabilités*. Hermann: Actes du Colloque de Genève sur la Théorie des Probabilités.

Gillies, D. (2000). Varieties of propensity. *British Journal for the Philosophy of Science, 51,* 807–835.

Gray, J. (2000). *The Hilbert challenge.* Oxford: Oxford University Press.

Hacking, I. (1965). *The logic of statistical inference.* Cambridge: Cambridge University Press.

Hájek, A. (1996). Fifteen arguments against finite frequentism. *Erkenntnis, 45*(2/3), 209–227.

Hájek, A. (2003a). Interpretations of probability. In *The Stanford encyclopedia of philosophy* (sec. 3.2); available at http://plato.stanford.edu/entries/probability-interpret/

Hájek, A. (2003b). What conditional probability could not be. *Synthese, 137,* 273–323.

Hájek, A. (2009). Fifteen arguments against hypothetical frequentism. *Erkenntnis, 70*(2), 211–235.

Hawkins, T. (2001). *Lebesgue's theory of integration: Its origins and development* (2nd ed.). Providence RI: American Mathematical Society.

Hendricks, V. F., Pedersen, S. A., & Jørgensen, K. F. (Eds.). (2001). *Probability theory: Philosophy, recent history and relations to science.* New York: Springer.

Heyde, C. C., Seneta, E., Crepel, P., Fienberg, S. E., & Gani, J. (Eds.). (2001). *Statisticians of the centuries.* New York: Springer.

Hitchcock, C. (2002). Probability and chance. In *The international encyclopedia of the social and behavioral sciences* (Vol. 18, pp. 12, 089–12,095). London: Elsevier.

Humphreys, P. (1985). Why propensities cannot be probabilities. *Philosophical Review, 94*(4), 557–570.

Jaynes, E. T. (1986). Bayesian methods: General background. In J. H. Justice (Ed.), *Maximum-entropy and Bayesian methods in applied statistics.* Cambridge: Cambridge University Press.

Jeffreys, H. (1961). *Theory of probability.* Oxford: Oxford University Press.

Keynes, J. (1921). *A treatise on probability.* New York: Macmillan.

Kolmogorov, A. (1933a). Sulla Determinazione Empirica di una Legge di Distribuzione. *Giornale dell' Istituto Italiano degli Attuari, 4,* 83.

Kolmogorov, A. (1968). Logical basis for information theory and probability theory. *IEEE Transactions on Information Theory, 14*(5), 662–664.

Laplace, P. S. (1774). Mémoire sur la Probabilité des Causes par les Événements. *Mémoires de l'Académie Royale des Sciences de Paris, 6,* 621–656.

Laplace, P. S. (1814). *Essai Philosophique sur les Probabilités.* Paris: Courcier.

Laplace, P. S. (1875). *Theorie Analytique des Probabilités.* Paris: Ve. Courcier.

Laplace, P. S. (1904). *Oeuvres Complètes de Laplace.* Paris: Gauthier-Villars.

Leibniz, G. V. (1989). *Philosophical essays.* Indianapolis: Hackett Publishing.

Lindley, D. V. (1965). *Introduction to probability and statistics from a Bayesian viewpoint.* Cambridge: Cambridge University Press.

Popper, K. (1935). *Logik der Forschung.* Vienna: Julius Springer Verlag (Trans. *The logic of scientific discovery.* London: Routledge (1959).).

Popper, K. (1957). The propensity interpretation of the calculus of probability, and the quantum theory. In S. Körner (Ed.), *Observation and interpretation* (pp. 65–70). London: Butterworth.

Popper, K. (1959). The propensity interpretation of probability. *British Journal for the Philosophy of Science, 10,* 25–42.

Popper, K. (1963). *Conjectures and refutations: The growth of scientific knowledge.* London: Routledge.

Popper, K. (1994). Contributions to the formal theory of probability. In P. W. Humphreys (Ed.), *Patrick Suppes: Scientific philosopher.* London: Kluwer Academic Publishers.

Raiffa, H., & Schlaifer, R. (1961). *Applied statistical decision theory.* Cambridge: Cambridge University Press.

Ramsey, F. P. (1931). Truth and probability. In R. B. Braithwaite (Ed.), *Foundations of mathematics and other logical essays*. New York: Kegan, Paul, Trench, Trubner, & Co. (Reprinted in Kyburg Jr., Smokler (Eds.), *Studies in subjective probability*. New York: Wiley (1964).).

Reichenbach, H. (1935). *Wahrscheinlichkeitslehre: eine Untersuchung über die logischen und mathematischen Grundlagen der Wahrscheinlichkeitsrechnung* (Trans. *The theory of probability: an inquiry into the logical and mathematical foundations of the calculus of probability*. Berkeley: University of California Press (1948).).

Savage, L. J. (1954). *The foundations of statistics*. New York: Wiley.

Shanks, N. (1993). Time and the propensity interpretation of probability. *Journal for General Philosophy of Science, 24*(2), 293–302.

von Mises, R. (1928). *Wahrscheinlichkeit, Statistik und Wahrheit*. Wien: Springer. (Trans. *Probability, statistics and truth*. Stockton, CA: Courier Dover (1981).).

von Mises, R. (1951). *Positivism: A study in human understanding*. Cambridge, MA: Harvard University Press.

von Mises, R. (1963/1964). *Selected Papers of Richard von Mises* (Selecta I and II). Providence, RI: American Mathematical Society.

Ville, J. (1939). *Étude Critique de la Notion de Collectif*. Paris: Gauthier-Villars.

Wald, A. (1939). Contributions to the theory of statistical estimation and testing: A hypotheses. *Annals of Mathematical Statistic, 10*, 299–326.

Walley, P. (1991). *Statistical reasoning with imprecise probabilities*. London: Chapman and Hall.

Wang, Z., & Klir, G. J. (2009). *Generalized measure theory*. New York: Springer.

Appendix B: Pluralist Works—A Partial Bibliography

Anderson, N. H. (2001). *Empirical direction in design and analysis*. New York: Psychology Press.

Ballentine, L. E. (2001). Interpretation of probability and quantum mechanics. In A. Khrennikov (Ed.), *Foundations of probability and physics* (Vol. XIII, pp. 71–85). Singapore: World Scientific.

Barnett, V. (1999). *Comparative statistical inference*. Chichester: Wiley.

Bayarri, M. J., & Berger, J. O. (2004). The interplay of Bayesian and frequentist analysis. *Statistical Science, 19*(1), 58–80.

Berger, J. O. (2003). Could Fisher, Jeffreys and Neyman have agreed on testing? *Statistical Science, 18*(1), 1–32.

Box, G. E. P. (1983). An apology for ecumenism in statistics. In G. E. P. Box, T. Leonard, C. F. Wu (Eds.), *Scientific inference, data analysis, and robustness* (pp. 51–84). New York: Academic Press.

Bulmer, M. G. (1979). *Principles of statistics*. Mineola, NY: Courier Dover Publications.

Casella, G., Berger, R. L. (1987). Reconciling Bayesian and frequentist evidence in one-sided testing problems. *Journal of the American Statistical Association, 82*, 106–111.

Chatfield, C. (2002). Confessions of a pragmatic statistician. *Journal of the Royal Statistical Society, Series D, 51*(1), 1–20.

Chatterjee, S. K. (2003). *Statistical thought: A perspective and history*. Oxford: Oxford University Press

Cohen, J. (1989). *An introduction to the philosophy of induction and probability*. Oxford: Oxford University Press.

Costantini, D. (1970). *Fondamenti del Calcolo delle Probabilità*. Milano: Feltrinelli Editore.

Cox, D. R. (1971). The choice between alternative ancillary statistics. *Journal of the Royal Statistical Society, Series B*, 33, 251–255.

Daston, L. (1995). *Classical probability in the enlightenment*. Princeton, NJ: Princeton University Press.

Good, I. J. (1965). *The estimation of probabilities. An essay on modern Bayesian methods*. (Research Monograph, No. 30) Cambridge, MA: MIT Press.

Greenland, S. (2010). Comment: The need for syncretism in applied statistics. *Statistical Science*, 25(2), 158–161.

Gulotta, B. (1962). Probabilità e Frequenza. *Rendiconti del Seminario Matematico di Messina, 5* (1960/1961), 64–92.

Kamlah, A. (1987). What can methodologists learn from history of probability. *Erkenntnis*, 26, 305–325.

Kardaun, O. J. W. F. (2005). *Classical methods of statistics: With applications in fusion-oriented*. Berlin: Springer.

Kendall, M. G., & Stuart, A. (1977). *The advanced theory of statistics*. New York: Macmillan.

Kyburg, H. J. Jr. (1974). *The logical foundations of statistical inference*. New York: Springer.

Lehner, P. E., & Laskey, K. B. (1996). An introduction to issues in higher order uncertainty. *IEEE Transactions on Systems, Man, and Cybernetics Part A, 26*(3), 289–293.

Lewis, D. (1971). A subjectivist's guide to objective chance. Berkeley: University of California Press (Reprinted with postscripts in Lewis, D. (1986). *Philosophical papers*, II. Oxford: Oxford University Press, pp. 83–132.).

Little, R. (2006). Calibrated Bayes: A Bayes/frequentist roadmap. *The American Statistician, 60*(3), 1–11.

Mellor, D. H. (1971). *The matter of chance*. Cambridge: Cambridge University Press.

Nagel, E. (1945). Probability and Non-demonstrative Inference. *Philosophy and Phenomenological Research, 5*, 485–507.

Oakes, M. W. (1986). *Statistical inference: A commentary for the social and behavioural sciences*. Chichester: Wiley.

Pearson, E. S. (1962). Some thoughts on statistical inference. *The Annals of Mathematical Statistics, 33*, 394–403.

Pliego, F. J. M., & Pérez, L. R. (2006). *Fundamentos de Probabilidad*. Thomson.

Poirier, D. J. (1995). *Intermediate statistics and econometrics: A comparative approach*. Cambridge, MA: MIT Press.

Rizzi, A. (1992). *Inferenza statistica*. Torino: UTET Università.

Rubin, D. B. (1984). Bayesianly justifiable and relevant frequency calculations for the applied statistician. *The Annals of Statistics, 12*(4), 1151–1172.

Salmon, W. C. (1967). *Foundations of scientific inference*. Pittsburgh: University of Pittsburgh Press.

Samaniego, F. J., & Reneau, D. M. (1994). Toward a reconciliation of the Bayesian and frequentist approach to point estimation. *Journal* of the *American* Statistical *Association, 84*, 947–957.

Smith, R. L., & Young, G. A. (2005). *Essentials of statistical inference*. Cambridge: Cambridge University Press.

Stauffer, H. B. (2007). *Contemporary Bayesian and frequentist statistical research methods for natural resource scientists*. Hoboken, NJ: Wiley-Interscience.

Tijms, H. (2011). *Understanding probability: Chance rules in everyday life*. Cambridge: Cambridge University Press.

van Fraassen, B. C. (1983). Calibration: A frequency justification for personal probabilities. In R. S. Cohen & L. Laudan (Eds.), *Physics, philosophy and psychoanalysis* (pp. 295–319). Oxford: Clarendon Press.

Williams, D. (2001). *Weighing the odds: A course in probability and statistics*. Cambridge: Cambridge University Press.

Wonnacott, R. (1990). *Introductory Statistics* (5th ed). New York: Wiley.

Yourgrau, W. (1961). Challenge to dualism. *The British Journal for the Philosophy of Science, 12*(46), 158–166.

Appendix C: Law of Large Numbers—A Proof

The proof of the law of large numbers turns out to be rather laborious and the majority of mathematicians resorts to intermediate achievements which pave the way toward the definitive demonstration. They mention one or more of the following results the *convergence of characteristic functions,* the *Chebyshev inequality,* the *Borel–Cantelli lemma,* the *Kolmogorov inequality* and the *Markov inequality.* The Chebyshev estimate may be replaced by the *Bernstein inequality* which is a more exact one but under more restrictive conditions.

Researchers are looking for a more agile method for proving the theorem such as Nasrollah Etemadi (1981) who has found a swift way to demonstrate the theorem.

This Appendix exhibits a demonstration of the LLN based on Borovkov (1998). The proof includes three main steps: the first step deals with the Chebyshev inequality, the second proves the Kolmogorov inequality and the third one focuses on the Bernoulli trials.

Chebyshev's Inequality

Suppose that the random variable ξ is distributed on positive half axis with a continuous time density $p_\xi(t)$ and mean $M\xi$ finite. From the mean definition we have

$$M\xi = \int_0^\infty t p_\xi(t)dt \geq \int_\varepsilon^\infty t p_\xi(t)dt \geq \varepsilon \int_\varepsilon^\infty p_\xi(t)dt = \varepsilon P(\xi \geq \varepsilon).$$

where ε is any positive real number. Then the following inequality—called *Bienome–Chebyshev inequality*—is true

$$P(\xi \geq \varepsilon) \leq \frac{M\xi}{\varepsilon}. \tag{11.1}$$

Now we may consider random variables which are independent and identically distributed (i.i.d.r.v.) $\xi_1, \xi_2, \xi_3, \ldots$ coincide by the distribution with the random variable ξ with the finite mean $a = M\xi_1$, and the finite variation $b = \mathrm{Var}\xi_1$. Let $a_n = (\xi_1 + \ldots + \xi_n)/n$, and $A_n = (a_n - a)^2$ then

$$MA_n = Vara_n = b/n.$$

We calculate the probability that a_n differs from a by at least ε, and then use the Chebyshev–Bienome inequality

$$P\left(A_n \geq \varepsilon^2\right) = P(|a_n - a| \geq \varepsilon) \leq \frac{b}{n\varepsilon^2}.$$

We obtain the following inequality usually called *Chebyshev inequality*

$$P(|a_n - a| \geq \varepsilon) \leq \frac{b}{n\varepsilon^2} \to 0, \quad n \to \infty \quad \text{for any } \varepsilon > 0. \qquad (11.2)$$

This result is proved for random variables ξ_1, ξ_2, ξ_3, ... with continuous density. But it is easy to extend this proof to random variables with general distribution.

Kolmogorv's Inequality

Suppose that i.i.d.r.v. ξ_1, ξ_2, ξ_3,... have the finite mean $a = M\xi_1$, and the finite variation $b = \mathrm{Var}\xi_1$. Let us denote $S_n = \xi_1 + \xi_2 + \cdots + \xi_n$ and designate $\tilde{S}_n = \max(|S_n|, \cdots |S_n|)$, we can write the Kolmogorov inequality as

$$P(\tilde{S}_n \geq \varepsilon) = P(\tilde{S}_n^2 \geq \varepsilon^2) \leq \frac{VarS_n}{\varepsilon^2}. \qquad (11.3)$$

It is useful to specify that the Chebyshev inequality may be rewritten as follows

$$P(|S_n| \geq \varepsilon) = P(S_n^2 \geq \varepsilon^2) \leq \frac{VarS_n}{\varepsilon^2}. \qquad (11.4)$$

So the Kolmogorov inequality is an enhancement of the Chebyshev inequality.
Proof Denote

$$\eta(\varepsilon) = \min(k : |S_k| \geq \varepsilon).$$

When

$$\sum_{k=1}^{\infty} I(\eta(\varepsilon) = k) = 1.$$

where $I(a)$ is the indicator function of the event A. Using the previous definitions we have

$$MS_n^2 = \sum_{k=1}^{\infty} MS_n^2 I(\eta(\varepsilon) = k) \geq \sum_{k=1}^{n} MS_n^2 I(\eta(\varepsilon) = k)$$

$$= \sum_{k=1}^{n} M((\xi_1 + \cdots + \xi_k) + (\xi_{k+1} + \cdots + \xi_n))^2 I(\eta(\varepsilon) = k)$$

$$\geq \sum_{k=1}^{n} M(\xi_1 + \cdots + \xi_k)^2 I(\eta(\varepsilon) = k) + 2\sum_{k=1}^{n} M(\xi_1 + \cdots + \xi_k)(\xi_{k+1} + \cdots + \xi_n) I(\eta(\varepsilon) = k).$$

It is obvious that

$$M(\xi_1 + \cdots + \xi_k)(\xi_{k+1} + \cdots + \xi_n) I(\eta(\varepsilon) = k) = 0.$$

Then the following holds

$$MS_n^2 \geq \sum_{k=1}^{n} M(\xi_1 + \cdots + \xi_k)^2 I(\eta(\varepsilon) = k)$$

$$\geq \sum_{k=1}^{n} M\varepsilon^2 I(\eta(\varepsilon) = k)$$

$$= \sum_{k=1}^{m} \varepsilon^2 P(\eta(\varepsilon) = k)$$

$$= \varepsilon^2 P(\eta(\varepsilon) \leq n)$$

$$= \varepsilon^2 P(\tilde{S}_n \geq \varepsilon).$$

Inequality (11.3) is proved.

The Strong Law of Large Numbers for Bernoulli Trials

In the first step, we formulate the strong law of large numbers as follows

$$\frac{S_n}{n} \to a \quad \text{as} \quad n \to \infty \text{ almost surely.} \tag{11.5}$$

where S_n/n is the relative frequency of successes in n Bernoulli trials.
Proof Without loss of generality it is possible to make the following almost surely equivalent formulations of this statement

$$\frac{S_n}{n} \to a, \ n \to \infty \Leftrightarrow$$

$$\forall \varepsilon > 0 \exists N : \forall n > N \left| \frac{S_n}{n} - a \right| \leq \varepsilon \Leftrightarrow$$

$$\forall \varepsilon > 0 \exists n : \sup_{k \geq n} \left| \frac{S_n}{n} - a \right| \leq \varepsilon \Leftrightarrow$$

$$\bigcup_{n \geq 1} \bigcap_{k \geq n} \left(\left| \frac{S_n}{n} - a \right| \leq \varepsilon \right) = \Omega.$$

Which leads to

$$\frac{S_n}{n} \to a, \ n \to \infty \Leftrightarrow$$

$$\bigcap_{n \geq 1} \bigcup_{k \geq n} \left(\left| \frac{S_n}{n} - a \right| \geq \varepsilon \right) = \varnothing.$$

where Ω is the complete event and $\Leftrightarrow$ is the empty event. The last relation may be rewritten in the following terms

$$P \left(\bigcap_{n \geq 1} \bigcup_{k \geq n} \left(\left| \frac{S_n}{n} - a \right| \geq \varepsilon \right) \right) = 0.$$

So to prove the strong law of large numbers it is sufficient to show that

$$P \left(\bigcap_{n=1}^{\infty} A_n \right) = 0, \quad A_n \left(\bigcup_{k \geq n} \left(\left| \frac{S_n}{n} - a \right| \geq \varepsilon \right) \right), \quad A_1 \supseteq A_2 \supseteq A_3 \ldots$$

Or

$$P(A_n) \to 0, \quad n \to \infty. \tag{11.6}$$

Without loss of generality it is possible to take $a = 0$ and Eq. (11.6) can be rewritten as follows

$$P \left(\sup_{k \geq n} \left| \frac{S_k}{k} \right| > \varepsilon \right) \to 0, \quad n \to \infty.$$

Substituting the superior term with the maximum

$$B_k = \left(\max_{2^{k-1} \leq j < 2^k} \left| \frac{S_j}{j} \right| > \varepsilon \right).$$

And introducing B_k into the previous limit

$$P \left(\bigcup_{k \geq n} B_k \right) \to 0, \quad n \to \infty.$$

The following holds true

$$P\left(\bigcup_{k \geq n} B_k\right) \leq \sum_{k \geq n} P(B_k)$$

$$= \sum_{k \geq n} P\left(\max_{2^{k-1} \leq j < 2^k} |S_j| > \varepsilon 2^{n-1}\right) \leq \sum_{k \geq n} P(\tilde{S}_{2^k} > \varepsilon 2^{k-1}).$$

Using Kolmogorov inequality (11.3) we obtain the following result that proves Eq. (11.5)

$$P\left(\sup_{k \geq n}\left|\frac{S_k}{k}\right| > \varepsilon\right) \leq 4 \sum_{k \geq n} \frac{VarS_{2^k}}{\varepsilon^2 2^{2k}} = 4 \sum_{k \geq n} \frac{2^k b}{\varepsilon^2 2^{2k}} = \frac{4b}{\varepsilon^2} \sum_{k \geq n} \to 0, n \to \infty.$$

In the second step, we assume that i.i.d.r.v. ξ_i; $i > 0$ characterize success or failure in Bernoulli trials and satisfy the following conditions

$$P(\xi_1 = 1) = p,$$
$$P(\xi_1 = 0) = q,$$
$$p > 0,$$
$$q > 0,$$
$$p + q = 1.$$

Because

$$M\xi_1 = a = p.$$

And

$$Var\xi_1 = b = pq.$$

Random variables ξ_i $(i > 0)$ satisfy the strong law of large numbers in the ensuing form

$$\frac{S_n}{n} \to p \quad \text{as} \quad n \to \infty \text{ almost surely.}$$

References

Borovkov, A. A. (1998). *Probability Theory*. Amsterdam: Gordon and Breach.
Etemadi, N. (1981). An elementary proof of the strong law of large numbers. *Z. Wahrsch. Verw. Gebiete, 55*(1), 119–122.

Index

P. Rocchi, *Janus-Faced Probability*, DOI: 10.1007/978-3-319-04861-1,
© Springer International Publishing Switzerland 2014

Printed in the United States
By Bookmasters